Civil Service Reform in Post-Communist Countries

AF859735

Leiden University Press

Cover design: Maedium, Utrecht

ISBN 978 90 8728 060 4
e-ISBN 978 90 4850 794 8
NUR 740

© A. Kotchegura / Leiden University Press, 2008

All rights reserved. Without limiting the rights under copyright reserved above, no part of this book may be reproduced, stored in or introduced into a retrieval system, or transmitted, in any form or by any means (electronic, mechanical, photocopying, recording or otherwise) without the written permission of both the copyright owner and the author of the book.

Civil Service Reform in Post-Communist Countries

The Case of the Russian Federation and the Czech Republic

PROEFSCHRIFT

ter verkrijging van
de graad van Doctor aan de Universiteit Leiden,
op gezag van de Rector Magnificus prof.mr. P.F. van der Heijden,
volgens besluit van het College voor Promoties
te verdedigen op woensdag 19 November 2008
klokke 15.00 uur

door

Alexander Kotchegura

geboren te Magdeburg,
in 1956

Promotiecommissie

Promotor:

Prof. A.J.G.M. Bekke

Co-promotor:

Dr. F.M. van der Meer

Referent:

Prof. J.C.N. Raadschelders (Oklahoma University)

Overige leden:

Jhr. Dr. F.K.M. van Nispen tot Pannerden (Erasmusuniversiteit Rotterdam)
Prof. Dr. M.R. Rutgers
Prof. Dr. Th.A.J. Toonen

Contents

Acknowledgements

The preparation and publication of this study would have not been possible without invaluable contributions made by a number of people. I would like to express my gratitude to many colleagues from the Department of Public Administration of the Faculty of Social and Behavioural Sciences of Leiden University, who shared their academic and scientific knowledge, skills and experience with me and provided recommendations on improving the work done.

I also thank colleagues from my home university who supported me throughout all the research and whose advice and assistance are highly appreciated.

This PhD research was a difficult but rewarding experience. My wife Julia, my daughter Ann and my son Vladislav, who was born while I was in the midst of my research activities, have been a source of encouragement and inspiration for me and I would like to express my gratitude for the support they provided.

A.K.
August 2008

CIVIL SERVICE REFORM IN POST-COMMUNIST COUNTRIES: THE CASE OF THE RUSSIAN FEDERATION AND THE CZECH REPUBLIC

"Every revolution evaporates and leaves behind only the slime of a new bureaucracy".

Franz Kafka

CHAPTER I: INTRODUCTION

1.1 Transformation of the State and Public Sector Reforms in Central and Eastern Europe

The late eighties and early nineties of the 20th century saw unprecedented changes in the political and economic systems of the Central and Eastern European (CEE) countries. Hardly anywhere else did so many countries undertake transformation of their basic political and economic institutions on such a scale. The retreat from one party rule, bureaucratic centralism and central planning towards a new system characterized by market economy, democratic institutions and developed civil society posed a formidable challenge for the nations in question.

The countries in transition faced the task of establishing the basics of a democratic society: in particular adopting new constitutions, transferring power to elected representatives, creating a multi-party system, establishing a favourable and competitive environment for free market operators, encouraging development of civil society organizations and promoting independence of mass media.

The task of such enormity necessitated a profound transformation of the state administration and the overall system of governance. This pertains in particular to the re-orientation of the civil service towards an impartial execution of policy mapped out by the political party in power, ensuring the rule of law and eliminating possibilities for arbitrary use of public authority. This also means introduction of appropriate internal and external control procedures and establishing permanent dialogue between public servants and citizens as well as maintaining strong accountability of the former to democratic institutions and civil society.

The formation of an efficient, cost effective, accountable, based on merit and corruption immune civil service turned out to be an extremely complicated task in virtually all post-communist countries. Notwithstanding progress in some areas the bulk of this work still lies ahead in the majority of CEE countries.

Although civil service reform issues were put high on the agenda of CEE governments, in many cases reform processes were slow, delayed or delivering other than planned results. By the mid 1990s state bureaucracy in many Central and Eastern European countries appeared to have been least of all affected in real terms by the transformation when compared to other public institutions and sectors of the economy (Hesse, 1998; Nunberg, 2000; Goetz, 1999; Larjavaara, 2001; Verheijen, 2002; Meyer-Sahling, 2002; Kotchegura, 1999 and 2004). Understanding that difficulties and problems of post-communist transition are to a considerable extent rooted in the essentially unreformed nature of the bureaucracies in the countries concerned started to grow among politicians, experts and international donor organizations in the second half of the 1990s.

1.2 Civil Service Reform: International and Regional Context

Civil service and administrative reforms appear to be in fashion nowadays. The latest wave of reforms has been launched first across a limited number of countries in the mid-1980s (Australia, New Zealand, Great Britain and the USA are most often cited) spreading then further to other countries and continents. Many prominent politicians and scholars argue that reform and modernisation of contemporary state bureaucracy should become a permanent non-stop process.

Obviously there are good reasons for the renewed interest in modernisation and transformation of state administration. The development of modern society places much greater demand on the quality, efficiency and orientation of the activities of civil servants. Much higher priority has been recently attached worldwide to the issues of civil servants' performance, transparency, accountability, client orientation, integrity and responsiveness.

In spite of certain commonalities inherent to the majority of reform processes it is possible to distinguish at least two distinct trends in the reform of public management in the industrially developed countries over the recent decade. Using the phraseology and rationale provided by Pollitt and Bouckaert we can refer to the New Public Management (NPM) and the New Weberian State (NWS) as major reform trends. According to the same authors the adherents of the NWS model give priority to the modernization of the Weberian tradition, whereas NPM proponents largely reject this tradition and promote approaches widely used in the private sector (Pollitt & Bouckaert, 2004: 46).

World-wide experience also demonstrates that reform of civil service systems is a complex process involving competing and conflicting interests and is subject to a multitude of exogenous and endogenous factors. "We are convinced that a conceptually identical, or at least very similar, reform develops differently in one national context as compared with another." (Pollitt & Bouckaert, 2004:39).

In Central and Eastern Europe transition to a new political, economic and social reality in the 1990s has been highly uneven and overall characterised by advances and setbacks in practically all countries in the region. The political and economic transformation challenged the capacity of the governments and administrative structures to manage the reform processes and react to a constantly changing environment.

The timing, contents, pace and outcomes of administrative reform efforts were unique in each country of the region. At the same time one can notice certain general tendencies and commonalities. To a considerable extent they stem from jointly shared historical legacies but there are also other factors. These similarities are above all manifested in the specific features of state bureaucracies in the CEE region – widespread clientelistic networks and abuse of public office, high politicisation, considerable policy making opportunities and weak enforcement of civil service legislation, to name just a few.

By the late 1990s transformation of the public administration system has been acknowledged to be a crucial element of the transition process in CEE states. The agenda of in-depth reform of the civil service in post-communist countries incorporates measures aimed at streamlining their institutional role; enhancing accountability and efficiency, transparency and responsiveness; enforcing political neutrality and strict adherence to the rule of law; introducing modern management techniques and effective anti-corruption strategy; and strengthening performance and client orientation.

Preparation for EU membership and subsequent accession has been a separate and quite significant factor that affected the political, economic and institutional development of many CEE countries. This factor has served as an additional stimulus for speeding up modernisation and reform of the civil services in the region.

1.3 Rationale for the Selection of Cases

Studying institutional change, in particular reforms of administrative systems, is essential for gaining full understanding of the process of transformation that has been taking place in Central and Eastern Europe since the late 1980s-early 1990s. Available evidence demonstrates that institutions have become influential actors in modern political systems and they define the framework within which politics takes place (see Chapter II for more information).

In recent years many scholars and researchers have increasingly pointed to the importance of studying the process of administrative transformation in Central and Eastern Europe as this offers "the most exciting prospect for comparativists: the study of de-bureaucratization in Eastern Europe" (Fried, 1990:339). In spite of generally high interest and demand for comparative research, "the number of genuinely comparative studies of public service systems remains small" (Raadschelders, 1998:160), "the study of civil service reform in Central and Eastern Europe is still in an early state of development" (Meyer-Sahling, 2002:5) and "questions referring to the timing, progress and outcomes of civil service reforms in Central and Eastern Europe are far from being answered." (ibid).

The overall objective of the present research is to explore the process of transformation of civil service systems in selected post-communist states. The selection of countries for the research was therefore confined to Central and Eastern Europe. Within the region the choice of the Russian Federation and the Czech Republic has been deliberate and is explained as much for their differences as for their similarities. In particular, the following points are worth noting:

- In the early 1990s in the atmosphere of overall fascination with market reforms in Central and Eastern Europe there was general lack of apprehension of the importance of creating a competent, effective and accountable state administration. Though quality of public management has been gradually recognized by the Russian and Czech politicians as a key factor in ensuring success of crucial reforms, this awareness has not been followed by sustained and well coordinated actions to design and implement an effective strategy of civil service reform. In essence a kind of contradiction has gradually grown in both Russia and the Czech Republic. The political and the economic systems of both countries have drastically changed but the way these countries had been managed administratively was largely retained. This contradiction appears to have influenced many political, economic and social processes in the countries concerned;

- Both Russia and the Czech Republic have gone through the so called "triple transformation" – a specific form of de-institutionalization in politics, economy and nationhood in the early 1990s and this combination differentiates them from the absolute majority of other post-communist countries;

- The Russian Federation and the Czech Republic stand out from the majority of other CEE countries in that their civil service reform was considerably delayed. One of the objectives that guided the author's research effort was intention to contribute to the identification of principal causes of such delays;

- In view of the considerable delay in start-up of civil service reforms in both countries it is of high academic and practical interest to explore how factors of *similarity* (e.g. the recent communist past manifested in the dominating ideology and mentality, prevailing administrative practices and attitudes, complicated economic situation, etc) and *variance* (e.g. size of the territory and population, availability of natural resources, constitutional arrangements, traditions and culture, external influence, etc) typical of each country affected the process of preparation and implementation of reform;

- The demand to improve governmental performance and to adjust state administration to the vastly changed role of the state and no less important to the new needs and expectations of the society has become more urgent in recent years and was made evident in significant developments in both countries. In the Czech Republic noted were the approval by the Czech Parliament of the Civil Service Act in 2002 and endorsement by the Czech Government of the Programme of Reform of Central State Administration in 2004. In the Russian Federation marked were the approval in 2002 by the Russian President of the Programme of Civil Service Reform for the years 2003- 2005 and adoption by the State Duma of the new Laws on Civil Service in 2003 and 2004;

- General shortage of studies devoted to various aspects of administrative transition in the researched countries and in particular actual absence of comprehensive comparative research, which considers reforms of the Russian and Czech civil services as its subject matter.

The appropriateness of these and other considerations and questions with regard to civil service reforms in Russia and the Czech Republic have been noted by observers and experts in the field: "It remains puzzling why a country (Czech Republic-A.K.) with good records in economic and political reforms and new member of the European Union has been unable for a long time to produce major progress in the area of civil service reform" (Meyer-Sahling, 2002:5).

In a similar way far reaching administrative reform launched in 2003 in Russia, in a country with deeply rooted traditions of "untamed bureaucracy", came somewhat as a surprise for most observers. The World Bank 2004 Russian Economic Report noted "In March 2004, the authorities announced – and immediately moved to implement – a blueprint for restructuring the Federal Government, which was far more radical than commentators or most public officials have been expecting" (World Bank, 2004:2).

What are the roots of these significant developments in both countries? Are the declared objectives and tasks of the reforms feasible? Are the reform plans being realized? What are the driving forces and constraints which either facilitate or complicate the advance of reform? What are current results of reform implementation? These and other questions arouse considerable interest and guided the author's effort throughout all the research work.

Although the programmes for large scale reform of core civil service appear to have been launched only recently in both Russia and the Czech Republic, it would be wrong to assume that no relevant developments took place in each country throughout the 1990s. In fact, preparation of the reform continued to a greater or lesser degree during all this period and was accompanied by debate, contradictions, conflicts, advances and setbacks, partial improvements and separate practical measures aimed at modernization of respective civil services. Analysis of these developments, together with comparison with the experience of other countries, may be helpful in avoiding difficulties and pitfalls in future and facilitating practical implementation of the reforms in Russia, the Czech Republic and other post-communist countries.

1.4 Objective, Research Questions and Structure of the Study

The focus of the research is on the identification, analysis and comparison of civil service reform initiatives and processes in the two selected countries starting from the early 1990s.

The principal research question is formulated as:

'What have been the experiences with the preparation, launching and implementation of national civil service reform in the Russian Federation and the Czech Republic following the fall of the communist regimes and which factors can explain the relative success or failure of the reform proposals'.

The following research sub-questions defined directions and contents of the thesis:

1) What were the main characteristics and prevailing features of each of the two civil service systems (Russian and Czech) before the reform was designed and launched?;
2) What are the dynamics of change and contents of reform of the Russian and Czech civil service systems?;
3) What are the essential factors that influenced the pace and current results of reform?;
4) How can the differences and similarities in the development and reform of the two civil services be compared and explained in order to define possible causes of similarities and variations?

In finding answers to the above mentioned principal questions of the research a number of related aspects of civil service reform preparation and implementation will need to be clarified, in particular: *a) Who or what initiates reform and why?; b) What are the dimensions of reform?; c) What is the extent of external influence?; d) What are the differences and similarities in launching and implementing reform?;*

The research also aims to contribute to the debate of why some post-communist countries made notable advance in reforming their state administrations while others have been delaying these reforms or putting them on hold.

The composition of the thesis is determined by its objectives and research questions as well as by the selected methodological approach. The *first chapter* introduces the reader into the topic, explains the selection of cases, defines the structure of the thesis and research questions. The *second chapter* provides an overview of the theoretical and methodological basis of research. The *third and the fourth_chapters* comprise an analysis of the Russian civil service system and reform of the Russian civil service respectively. The *fifth and the sixth chapters* offer a similar analysis of the Czech civil service system and reform of the Czech civil service. Country specific chapters are structured to a common format to facilitate comparison. The *seventh chapter* compares selected features of the Russian and Czech civil service systems in line with the internationally recognized protocol for comparative studies. The main objective of the *eighth chapter* is to draw together the findings of the previous chapters, to carry out comparison and to make on that basis certain conclusions. The *ninth chapter* summarizes principal results and inferences of the research, matching them against original assumptions and hypotheses as well as against developed theoretical approaches and propositions. Each chapter begins with a short introduction and ends with a brief conclusion.

The analysis focuses both on a 10-year period since the fall of communist regimes in Central and Eastern Europe and on recent events linked to the reform of civil service systems in both Russia and the Czech Republic to the extent latest information is available.

1.5 Sources of Information/Data and Related Constraints

Apart from traditional sources of data – various studies, monographs, analytical reports on the status of public administration and civil service, relevant legislation and other normative regulation - the author also used interviews with the Russian and Czech high and middle ranking civil servants, politicians and experts. The interviews were carried out by independent foreign and local consultants and also by the author at conferences, workshops, at specially organized meetings in Russia and during his numerous visits to the Czech Republic.

For the purpose of this research the findings of sociological studies, surveys and opinion polls have also been widely used. The author conducted a special standardized structured interview of Russian civil servants similar to the one made in the Czech Republic by a group of scholars within the framework of the Civil Service Research Project funded by CERGE-EI (a think tank affiliated to Charles University, Prague).

In spite of recent improvements, there still exists a shortage of reliable statistical data on the civil services in both countries. As a rule, information available from the various sources is incomplete and conflicting. Therefore in many cases presented data serve the purpose of outlining only general dimensions and highlighting prominent cases. "...Despite the heroic efforts of organisations such as the OECD, there are still many issues for which relevant data are either non existent, of doubtful reliability, or of doubtful comparability" (Pollitt & Bouckaert, 2004:21).

One of the difficulties faced by the author was the fact that preparation of civil service reforms in Russia and the Czech Republic took much more time than originally anticipated. The actual start of reforms occurred several years after this research was launched. The author considered that analysis of only the preparatory stage of reforms would be detrimental to the overall quality of the study and prolonged the work sufficient to see the initial results of the reforms to emerge.

Another constraint consisted in different, often contradictory understanding of the terms "public administration reform" and "civil service reform" in the countries concerned. For example, throughout the 1990s public administration reform in the Czech Republic was perceived exclusively as reform of the regional and local government.

For pragmatic reasons the analysis was focused mainly on the central government level of the Russian and Czech civil services, although attention to various levels of regional government was drawn whenever it was considered to be appropriate. Major findings and inferences of the research were included in various publications written by the author in course of preparation of the thesis.

In the absence of a universally acknowledged methodological approach to the comparative analysis of civil service reforms much attention has been paid to the identification of the proper theoretical basis of the research and the selection of appropriate methodological tools. This search led the author to join an international team of researchers from various countries led by academics and experts from Leiden (the Netherlands) and Indiana (USA) Universities. This team/consortium designed a Protocol for comparative studies of national civil service systems. The protocol constitutes a common methodological platform (framework) for making meaningful comparisons of these systems across different nations. The key elements of this framework were presented in the publication edited by Bekke, Perry and Toonen "Civil Service Systems in Comparative Perspective" (1996).

The members of the research consortium initiated a number of comparative studies resulting in a series of publications (issued in the late 1990s to the early 2000s) on the comparison of civil service systems in Northern America, Western Europe, Central and Eastern Europe and Asia. The author took active part in the preparation and co-edited one of these publications "Civil Service Systems in Central and Eastern Europe". Involvement in this common undertaking served as a source of inspiration for continuation of these activities and encouraged the author to start writing this book. Therefore, this study may be viewed as continuation of a broader research project.

The author hopes that apart from pure academic interest, the presented work will also have certain practical significance in particular for those who design and manage civil service reforms in the post-communist countries and elsewhere.

CHAPTER II
COMPARISON OF CIVIL SERVICE REFORMS: ISSUES OF THEORY AND METHODOLOGY

2.1 Introduction

This chapter provides an overview of the theoretical basis of the research and defines methodological instruments which are applied to reach its objectives. The first part of the chapter is devoted to the analysis of modern theoretical and methodological approaches to comparative public administration. It starts with the review of the current stage of development of the administrative science. This is followed by the study of methodological concepts and tools that are applied in the research of public administration.

We proceed further with the examination of the peculiarities of comparative studies, focusing on constraints and difficulties related to such studies in particular in public administration. We explain the choice of institutional concept and approach as the principal methodological tool. The relevance of institutional analysis is emphasized, underlining the significant role of formal and informal institutions (rules, norms, traditions, behavioral patterns) in enabling or constraining policy formulation and implementation. Subsequently, we highlight the usefulness of the Protocol for comparative studies of national civil service systems (designed in the mid 1990s by a group of mainly Dutch and American scholars) as a methodological platform and principal framework for our comparative research and determine a list of potential variables.

The second part of the chapter starts with a summary of contemporary views on civil service reform. Existing opinions on reform dimensions and outcomes are reviewed and administrative reform processes in the Central and Eastern European region are considered in terms of their main specific features. Then we outline a rationale for the selection of key topics and structure of the study and explain the chosen model of analysis.

The chapter ends with a list of definitions of basic terms widely used in the research and a number of hypotheses and assertions stemming from the preliminary analysis.

2.2 Theoretical and Methodological Approaches in Comparative Public Administration

2.2.1 Administrative Science – Status and Prospects

The overall objective of this research is to study and compare civil service reforms in two selected countries. Setting this objective does not automatically mean acknowledgment of the existence of developed civil service systems in the above mentioned countries. Rather it encourages search for arguments that would either confirm or reject this assumption.

Civil service is a complex and multifunctional social institution and therefore can be studied from the point of view of various sciences - political science, sociology, law, psychology, history, management science etc. depending on a particular objective and subject of research as well as viewpoint taken.

However it is widely acknowledged that a comprehensive study of civil service systems is carried out primarily within the framework of a scientific discipline named *public administration science* or *administrative science.* The maturity of this science is sometimes questioned. This can be explained by failure so far to develop a sound conceptual framework accommodating the diversity of phenomena related to present day administrative process and transformation, and elaborate effective theoretical approaches for empirical country based or cross national analysis of administrative systems. Hence one may find statements that traditional public administration could not withstand new challenges and therefore "has been discredited theoretically and practically" (Hughes, 1994: 15). This led some authors to claim that public administration science experiences an 'identity crisis', the field lacks 'discipline'..." Rutgers points out that there are scholars who argue that "in order to become a respected field or discipline, the study of public administration should (re)gain an underlying coherence by means of a shared, generally accepted unifying paradigm" (Rutgers, 1998:8). In connection with this "a universal theory of public management will have to wait" (Peters, 1996: 19).

At the same time there exists a view that public administration will or should never be an integrated discipline. Reference is often made to the difficulty of establishing a standard paradigm "in a field that continued to value diversity rather than orthodoxy" (Heady, 1998: 34). Others reckon that the best way for administrative science to be useful for both academia and practitioners is "to develop differentiated integration" (Raadschelders, Rutgers, 2001: 15). In other words public administration should be perceived as "a multi-or interdisciplinary study that aims for differentiated integration of knowledge and that serves to provide a comprehensive view of government for all those that labor in the real world..." (Raadschelders, Rutgers, 2001: 3).

Looking at the problem from somewhat different angle it is difficult not to agree with the same authors that "the study of Public Administration is healthy, for it does not turn away from continuously raising fundamental questions about its identity" (Raadschelders, Rutgers, 2001: 2).

2.2.2 Developing Methodological Approaches in Administrative Research

Research methodology, broadly interpreted, encompasses various stages of the research process, and may be defined as the "system of explicit rules and procedures upon which research is based and against which claims for knowledge are evaluated" (Nachmias & Nachmias, 1982: 15).

There have been numerous attempts to develop a methodological and theoretical framework for public administration research. Some of these attempts are considered to have made notable advances in separate directions, but none is acknowledged as having a universal value. Diversity and pluralism of applied research concepts continues to be a typical feature of administrative science.

Viewed from a historical perspective the development of research in public administration, at least in the recent four decades, is often presented as a shift from a 'traditional' approach (focus on organisation) to 'neo-organisational' approach (with more attention to policy analysis and decision-making) and then to 'neo-institutional' approach (with considerable accent on path dependency and role of institutional setting).

In view of existing difficulties in creating a theoretical framework for public administration research, the idea of "searching at this stage for thicker and more useful descriptive statements about systems with the hope of producing first some descriptive generalizations and then perhaps science" (Peters, 1996: 33) appears sensible. Ideally these generalizations should be practice and comparison oriented i.e. they should give us some insight into "what works and what is best" (Peters, 1996: 33).

Furthermore, "...studying processes of change in several, individual countries in order to discover similarities and differences" may "give us elements for basic theory - building" (Bekke, 1999: 12). It may turn out that instead of a single, general theory on civil service systems on a high level of abstraction we will need "...several theories either directed to special aspects or single basic problems or built for separated collections of countries of the same kind" (Bekke, 1999: 12).

The appropriateness of going beyond structure and elements of civil service systems in question to incorporate their historical evolution and interface with the political system and other social systems and institutions has recently become more obvious. "Clearly there is something to be gained by a particular focus on administration, but we also need something of the big picture of the total system of politics and government" (Peters,1996: 29).

In connection with this Bekke stresses that "comparative research on civil service systems needs a strong link with both political theory and theories of societal structure and development" (Bekke, 1999: 11). For understanding why these differences and similarities exist, "we need to search for explanatory factors, not only directly given from within civil service systems as such, but also - and probably mainly - derived from the political and societal environment of the civil service systems in question" (ibid).

Speaking about the specific research methods in studying the development and transformation of civil service systems Bekke concludes "... we have to work mainly by inductive methodology, that is by studying processes of change in several, individual countries in order to discover similarities and differences" (Bekke, 1999: 11).

The inductive method relies heavily on descriptive data about actual situations and behaviour. It presupposes advance from specific observation to identify differences and commonalities leading to explanations and interpretations. The process concludes with generalizations explaining relationships among units observed.

Similarly Rutgers points to the existence of two methodological approaches in comparative research - quantitative and interpretative (Rutgers, 2004: 153).

The present research rests largely within the boundaries of macro-level of comparative analysis where our attention is focused primarily on the connectedness of administrative institutions to other political and social institutions in society. Micro-level analysis where the principal accent is placed on the individual public servant is applied occasionally wherever appropriate.

One of the frequently used methods of conducting specific comparative analysis of administrative systems pre-supposes application of typologies. The following typologies are usually singled out:

- *Ideal types (Weber's ideal type model often taken as a starting point);*
- *Configuration analysis (Heady, Morgan);*
- *Theoretical case study (Peters);*

The development of typologies (ideal types) may be helpful in setting "a standard against which the real world can be compared" (Peters, 1996: 29). Heady defines comparative studies of public bureaucracies based to a considerable extent on Max Weber's ideal-type model of bureaucracy as "the most utilized and the most frequently preferred alternative among the available frameworks for study..." (Heady, 1998: 34). However, many scholars share the view that this method does not help to identify the dynamics of change in social systems.

One of the well known typological frameworks for the comparative cross-country analysis of civil service systems was designed by Heady. In essence he developed a conceptual configuration that seeks to identify commonly occurring clusters of civil service system attributes. Heady introduced three sets of criteria - relation to the political regime, focus for personnel management and qualification requirements as central for defining the overall character of the system. Heady admitted also that he did not consider his set of configurations either ultimate or comprehensive. In our view, the research built on typologies and ideal types can be complimentary to our mainstream methodological approach.

2.2.3 Comparative Studies in Public Administration

Comparative research is one of the most general and commonly used strategies in the social sciences. “Comparison is the only laboratory open to most social scientists…” (Peters, 1996: 15). It involves examination of several similar units of analysis, taken as a rule in various environmental and historical contexts, in order to identify common and unique features inherent to them.

As a rule the comparative method enables:

- to formulate and test hypotheses;
- to improve classifications;
- to prescribe the future.

By focusing on patterns of administrative activities and the characteristics of the systems performing them, comparative research extends our understanding of conditions conducive to successful or unsuccessful administrative performance. It is not surprising, therefore that “the comparative approach is inherently inclined to center on administrative reform and capacity building” (Jreisat, 2002: 2).

The authors of ‘Civil Service Systems in Comparative Perspective’ single out three basic objectives of comparative research on civil service systems – 1) identification of key concepts, of relations among concepts, and of the underlying logic or dynamic of the relations; 2) development of better understanding of historical evolution of civil service systems; 3) enhancement of policymakers’ capacity to design civil service systems (Bekke, Perry, Toonen, eds. 1996: 7-8).

In spite of generally high interest and demand for comparative research, "the number of genuinely comparative studies of public service systems remains small" (Raadschelders, 1998: 160). To a considerable extent such a situation can be explained by the difficulties and problems encountered by scholars and practitioners in undertaking comparative research. They appear to face several problems and challenges. Among those difficulties (some already mentioned above) are:

a) lack of reliable and more or less precise data;
b) absence of universal theoretical approach;
c) still limited number of genuine comparative studies and a generally fragmented nature of their findings.

With regard to the variety of intellectual problems that comparative public administration faces, Peters writes that "we have not had the data bases nor the agreed upon conceptualisations necessary for more empirical work..." (Peters, 1996: 16). Among other constraints faced by comparative researchers he distinguishes (ibid):

a) absence of a single universally acknowledged agreement as to the meaning and contents of some key term and concepts, for instance "bureaucracy";
b) lack of accepted and easily operationalizable dependent variables;
c) problems associated with levels of analysis.

One of the difficulties of quantitative approaches in comparative research is that data on civil service systems are usually "dirty" and incomplete (McGregor & Solano, 1996: 43). Heady emphasizes that "... a sweeping comparison across national boundaries requires some organizing concept to avoid burial under an avalanche of data about a multitude of diverse administrative systems". (Heady, 1984: 60). In his view bureaucracy provides such an organizing concept.

Another problem of comparative studies is that of "translation" singled out in particular by Rutgers. He argues that in order to ensure relevant comparison the researcher can either try to understand and capture the authentic or local meaning of social reality in another place, or may develop his own concepts to apply as criteria of this reality (Rutgers, 2004).

In spite of all these difficulties we can witness a growing recognition that comparative public administration is a distinct field of research (Heady 1998, Drengsgaard 2003, Pollitt & Bouchaert, 2004).

Summing up that "comparative empirical research on civil service systems requires methodological pluralism" we concur with one of the conclusions made in the book "Civil Service Systems in Comparative Perspective" that because of proven "...embeddedness of civil service systems within broader administrative, political and social systems... close observation of behaviour within context rather than survey analysis would be an appropriate research strategy" (Bekke, Perry, Toonen eds. 1996: 326).

The same method but named differently is advocated by Hague who singles out three main techniques in the comparative research applied in particular by political scientists: case studies, statistical analysis and focused comparisons. The latter (in essence *close observation of behavior* – A.K.) "...has the strengths of both the other techniques but the weaknesses of neither."(Hague, 1992: 39).

In order to proceed further in our comparative research we need to adopt a framework that would allow systematic comparison between the two civil service systems and help deal with most of the difficulties linked with comparative studies (referred to above). For that we turn to the institutional theory and approach and try to determine its appropriateness for our study.

2.2.4 Institutional Concept and Institutional Approach

Recent decades have witnessed broad acceptance of the institutional approach and method as principal tools in empirical research (including comparative) and theory building in the administrative science. The current period of development of public administration theory is therefore often referred to as a period of *'neo-institutionalism'*.

It is widely acknowledged that the dynamics of institutional and administrative transformation are linked to changes in the political, economic, social and legal environments in which public institutions operate and on whose material and non-material inputs they largely depend. "In many ways, civil service systems are shaped and stabilized by formal or informal norms and values, historically developed and adapted to particular societal experiences and beliefs" (Bekke, 1999: 4).

Proceeding from the name of the concept one may conclude that its primary focus is on institutions. More precisely, the concept gives at least as much attention to the institutional foundations of administrative developments as to individual action. The present day "institutionalism" is "new" because it provides a broader vision of institutions which includes apart from organisations proper also ideas, values and norms. The "new" approach overcomes opposition between institutional and actor-centred explanations and rather aims at defining the role and "weight" of institutions and actors in explaining administrative changes.

In essence the institutional approach regards existing institutional configurations as independent explanatory variables in the analysis of political processes and institutional development. Hence, the process of institutional change is considered to be to a large extent affected by endogenous institutional dynamics of the existing institutional architecture.

As far as study of administrative changes is concerned Pollitt and Bouckaert consider that "in the field of public management reform the broader forces of economics and politics are almost always mediated through networks of institutions. The specific characteristics of these networks, and of the individual institutions that compose them, frequently have a profound shaping effect upon what actually happens during the course of reform, and therefore upon the final results and outcomes of the change process " (Pollitt & Bouckaert, 2004: 23).

Another approach, sometimes termed "agency based", defines human action as the main explanatory factor in institutional change, while institutions are viewed as intervening rather than determining variables. Pollitt and Bouckaert emphasize "we acknowledge the frequent constraining importance of path dependency, but also repeatedly stress the role of agency and choice" (ibid).

In view of this some authors (e.g. Goetz) argue that application of neo-institutional approaches for understanding post-communist governance has some limitations. Goetz points to "typicality of 'actor centered institutionalism' in post-communist governance" where he asserts "individuals often come first, institutions second". In the centre of this approach he places the process of elite decision making (Goetz, 1999: 26).

Occasionally this may be true, particularly at times when various institutions have not yet reached a sufficient level of maturity. However, it is our belief that even in societies undergoing drastic transformation, policies of change or conservation are seriously institutions bound. Politicians and decision makers, who consistently pursue policies that are in conflict with requirements imposed by the existing institutional framework, sooner or later have to either adapt or go. "Path dependency limits the potential role of leadership as it "carves" the institutional niche, which becomes especially clear when changing conditions demand a change of course" (Boin, Christensen, 2004: 9).

The same authors continue: "The path dependent nature of institutionalization is reinforced by the strong influence that institutionalized organizations can (and usually do) exert on individuals. People are, in principle, free to choose the public organization they want to work for, but once they are part of an institution they will feel a strong socializing pressure not only to learn and act within the formal structures, rules and roles but also to participate in creating or internalizing the informal norms and values" (ibid).

There have been also attempts to combine various methodological approaches, or at least to reduce their actual or virtual differences. For example, Toonen proposed to adapt public choice, network analysis and institutional analysis. For that he made use of a meta-theoretical framework developed by Kaser and Ostrom in order to identify different approaches and forms of institutional analysis. This framework presupposes the existence of three zones of actions; (1) operational choice, (2) collective choice and (3) constitutional choice. This differentiation was further developed and directly applied to the study of civil service systems in the publication "Civil Service Systems in Comparative Perspective" edited by Bekke, Perry, Toonen (1996). Toonen asserts that this model "...represents an organizational frame for arranging various theoretical ways of looking at a given subject matter" (Toonen, 1998: 235).

As the contributors to the above mentioned publication emphasize, and we concur, the institutional perspective presupposes acknowledgment of the persistence of rules, roles, behavioral patterns over time (Bekke, Perry, Toonen, eds. 1996:3). In other words it assigns an important role to the concept of *path dependency*. The contents of the latter term is variously interpreted, but many agree that it embraces a comparative historical approach. For instance Raadschelders views the *path dependency* concept as the one providing "...the necessary dynamic angle to an otherwise static analysis of institutions". According to Raadschelders the term *path dependency* "captures ... (1) the historical notions of continuity, diversity and change; (2) the historians' sensitivity to time and context; (3) the social scientists' preference for modelling and prediction" (Raadschelders, 1998: 570). Peters in his turn points out that methods of effective research on civil service systems should draw more from anthropology than from conventional social science (Peters, 1996:326).

Central to the institutional approach and broadly used in this context is the concept of *institutionalisation* - meaning formation, development, transformation and reform of civil service systems. "For many, institutional analysis in social sciences implies the notion of systems ruled by norms, culture and traditions rather than by organisational goals and rational perspectives." (Bekke, 1999: 3).

For the purpose of this research, the term 'institutionalisation' denotes the dynamics of the systemic change that is firmly 'embedded' in an environment of social institutions (Bekke 1999: 14). The infrastructure of social institutions consists of particular institutions created in the past. This institutional environment restricts the 'portfolio' of possible decisions that political actors can make, whilst it also empowers them to make certain decisions.

Administrative science literature, empirical studies and experts' opinion provide sufficient evidence that success or failure of reform strategies in industrially developed countries is largely determined by a combination of institutional and leadership factors.

In the political and administrative science the issue of "whether the primary explanation for the governmental performance is to be found in organizations and structures or in people" (Aberbach & Rockman, 1992:145) has been long debated. Recently there appeared a tendency to take a compromise perspective upon this question. The same authors assert "We cannot solve the organizations versus people problem other than to recognize the importance of each, and their interdependence" (ibid).

2.2.5 Protocol for Comparative Studies

In the mid 1990s an international team of researchers from various countries led by academics and experts from Leiden (the Netherlands) and Indiana (USA) Universities designed a Protocol (guidelines) for comparative studies of national civil service systems. The Protocol constituted a common methodological platform (framework) for making meaningful comparisons of these systems across a number of nations. The key elements of this framework were later presented in the publication edited by Bekke, Perry and Toonen "Civil Service Systems in Comparative Perspective".

The protocol regards civil service systems above all from the two main perspectives: *performance and legitimacy.* The general dissatisfaction with the state of affairs regarding these two aspects was the main driver for launching civil service reform in many countries in recent years. As the authors of the protocol correctly emphasize – no matter what type of reform has been implemented or attempted, and no matter in what political, economic and social context, civil service reform in general aims at improving performance of the civil service and the legitimacy of government action.

In compliance with institutional concept and principles of institutional analysis the protocol aimed to find answers to the following main questions:

- what social and legal rules determine the character of a civil service system;
- what is the origin of these rules;
- why the persist or change

To a considerable extent the research strategy and framework of this protocol determines the overall methodological approach to our study. The protocol serves as the base, which has been complemented and modified further for the purposes of our research. Our selection of topics for comparative analysis largely draws upon the contents and guidelines of this research protocol. The protocol defined the following topics, information on which needs to be collected and suitably processed to ensure appropriate analysis of single country cases:

Development of the Civil Service (history) with particular emphasis on the impact of changes in environment (political, economic, cultural, demographic and other) on the development of state administration;

Internal Labour Market with special focus on identifying and understanding the rules which govern issues of civil servants' entry, conditions of employment, reward mechanism, re-training etc;

Representativeness with special attention to the extent to which particular groups of civil servants represent various categories of the population and act as advocates of various interest groups;

Politicisation with particular stress laid upon interdependence existing between the political and administrative systems and degree of "clientelism" of civil servants in relation to the political elite and various lobbies;

Public Opinion with particular accent on the prevailing general perception of the civil service and civil servants in the society;

Reform and Diffusion - this aspect is central to our research and therefore deserves more attention. The Protocol for comparative studies of national civil service systems draws attention to the following three interrelated questions:

- *Who or what initiates reform and why? - with a particular emphasis on;*
  Which is the dominant tradition (governance or management)?
  Which is the category of reform (budgetary/financial, structural, procedural/technical, relational)?
- *What is the content of reform? - with a particular emphasis on;*
  What are the specific objectives and outputs of reform that should be achieved?
  How are the objectives and outputs planned to be achieved?
- *Extent of external influence? - with a particular emphasis on;*
  Borrowing of specific concept
  Borrowing of broad direction approach

Apart from the mentioned above publication edited by Bekke, Perry and Toonen our study draws from contributions made by other prominent scholars in the field. In particular, "Public Management Reform. A Comparative Analysis" by Pollitt and Bouckaert deserves in our opinion special attention in view of its high relevance to the objectives, substance and direction of our research.

2.2.6 Significance of Historical Legacy, Traditions, Behavioral Patterns and Political Setting

Institutional analysis highlights the importance of interfaces between the strategies of the actors and the fundamental rules and norms that constrain or enable policy formulation and implementation. As we stated earlier institutions define and limit the set of choices available to policy-makers. However, this preposition does not rule out such factors as individual or collective human action.

As far as institutions are concerned quite significant for our study are traditions and norms that have an impact on the contents and dynamics of changes within institutions and their environment. Heady in connection with this writes that the behavioral tendencies in bureaucracies "are extremely significant and should be primary objects for analysis and comparison" (Heady, 2001:75).

The existing norms, behaviour patterns and traditions do not come from nowhere. They are the product of historical development. Institutions, and hence civil service systems, may be regarded as "...a more or less coherent and enduring whole made up of norms and values that have developed and persisted over time". (Raadschelders and Rutgers, 1996: 68). Therefore it is hard to disagree with a view that "...historical perspective should be a standard element in all of the sub-fields in public administration" (Raadschelders, 2003: 162).

As we outlined above one of the best tools for "catching" the "right" historical perspective is the concept "path dependency". Applied to the study of administrative institutions, the notion of path dependency helps 'explaining patterns of institutional emergence, persistence and change' (Pierson 2000: 256). For many years the path dependency concept was associated mainly with the study of resistance of institutions to change and persisting continuity within policy formulation.

Our understanding of the path dependency concept has been recently expanded, in particular by Raadschelders. He argues that tradition thus far has been perceived mainly as inheritance, whereas much less attention has been given to the influence of tradition as creation or invention (Raadschelders, 2007:9). In other words tradition, in his view, may be regarded not only as a force of conservation, but also as a source of innovation. "It might very well be that tradition stifles change in one case, while, at the same time, it leads to change in another" (Raadschelders, 2007: 10).

No less important for our study are relationships between an administrative system and politics. Jones and Kettl observe "Politics lies at the core of management reform, not vice versa". They suggest that we need to pay much more attention to the interaction of political/electoral institutions and administrative institutions and reforms (Jones & Kettl, 2004:463). Pollitt and Bouckaert argue that "public management is always a part of the broader agenda of public governance". They advocate the idea of "integrating the study of management change with an analysis of political systems and contexts" (Pollitt & Bouckaert, 2004:2).

In general it can be said that "the civil service system, as a system, shares some of the properties of the government and political regime as a whole, and that the government and political regime in turn share some attributes of the surrounding society, economy, and culture" (Peters, 1996:21). Hence, "...comparative public administration is linked closely to the study of comparative politics, and must start from the base provided by recent and current developments in the comparative study of whole political systems" (Heady, 2001:7).

Existing political and administrative systems exert significant influence on the process of reform and change, in particular on the adoption of certain reform ideas and strategies and feasibility of achieving reform objectives. "It is in these linkages and relationships between administrative systems and their contexts that we may be able to identify key factors that explain why there has been no great administrative reform accomplishments in many developing countries despite repeated attempts" (Jreisat, 2002:66).

In this connection the authors of a recent study argue that "CSS (civil service system – A.K.) needs to be considered in relation to its immediate institutional/organizational environment: the political system and its office holders" (Raadschelders, Meer & Toonen, 2007).

Taking into consideration potential factors which may affect administrative change, Pollitt and Bouckaert identify five key elements that form the basis for comparing administrative systems and reforms:

1. The state structure (including constitution);
2. The nature of executive government at the central level (includes the nature of the political system);
3. Relationship between political executives and top civil servants;
4. The dominant administrative culture;
5. Diversity among the main channels through which the ideas that fuel public management reform emerge.

Subsequently they define some basic dimensions of these key elements. With regard to: (1) *state structure* these dimensions are vertical dispersion of authority and horizontal coordination at central government level; (2) *nature of executive government* at central level the range is between 'majoritarian' and 'consensual'; (3) *minister/mandarin relations* these are 'kind of bargain or deal' existing between top politicians and top civil servants and the 'extent of politicization of top bureaucrats'; (4) *administrative culture* consideration is given to two models - 'Rechtsstaat' and 'Anglo-Saxon' - plus intermediate models; (5) *sources of advice* concerning contents and strategy of reform, the variance may be in the range and diversity of these sources (Pollitt & Bouckaert, 2004:40).

As far as *scenarios of reform or reform strategy* are concerned they should comprise at least three elements: a) a vision of the desired future; b) analysis of the existing situation; and c) chain of measures or events in the direction of the desired future. The authors also point out that "fully worked out scenarios, with each of the three main elements clearly analyzed and described, are definitely the exception rather than the rule in public management reform" (Pollitt & Bouckaert, 2004:66).

Pollitt and Bouckaert also define three aspects of reform implementation:

- *the extent to which reform has been a top down or bottom up experience;*
- *the extent to which new organizations and structures have been created specifically to advance reform;*
- *the intensity of reform, that is, have governments barged ahead, trampling opposition underfoot, or have they tiptoed delicately, consulting and cooperating with the other stakeholders (such as public service unions) as they go? (*(Pollitt & Bouckaert, 2004:94).

Proceeding from the overview made above and the adopted institutional approach we can make preliminary conclusions set out in the following section as far as structure, priority topics and variables of the research are concerned.

2.2.7 Defining Variables and Structure of the Study

Taking into account the above overview and the guidelines contained in the Protocol for comparative studies we may conclude that important for us variables should be linked to the political, economic, social and legal environment of a civil service system, existing institutional arrangements (in their retrospective development) and historical legacy. Within these areas our special attention is drawn to rules, norms and traditions, as well as to political-administrative relations.

Therefore, a number of key variables could be identified in exploring such crucial characteristics as *civil service reform capacity*. The latter reflects in our view the prime parameter of the subject of the study and will help answer the key research questions and sub-questions - "...Whatever type of reform may be desired, not every country has an equal capacity to implement new arrangements in a coherent, broad scope way" (Manning & Parrison, 2003:4).

Civil service reform capacity can be defined as *the ability and potential to launch and successfully implement civil service reform.*

This capacity is largely determined by: a) existing formal and informal institutions and type of relationship between them, which can either constrain or facilitate reforms (including rules, norms, traditions, values and patterns of administrative behavior); b) collective or individual human action (willingness and readiness of the key agents – policy makers and stakeholders - to undertake action in the desired direction); and c) resources available (funds and expertise).

Study of civil service reform capacity may provide information on assessing existing potential for reform but is generally less helpful in assessing actual implementation of reform.

Rational judgments concerning the progress of reform implementation and the variables affecting this process can be made if *reform results are matched against initial plans and intentions.*

Finally, the degree of similarity and variance in the design and implementation of civil service reforms and subsequent conclusions as to the causes of their (relative) success or failure can be derived from the *comparison proper* of civil service reforms' capacities, the reforms' contents (dimensions), up-to-date results, related institutional settings (framework), internal or external pressure and leadership factor.

In this connection we define our *model of analysis* as consistent exploration of capacities for civil service reform in each of the countries concerned, followed by the analysis of the extent to which the reform plans and intentions have been realized, and then finalized by the comparison of similarities and differences in reform preparation and implementation, which will allow define explanatory factors and answer research question and sub-questions.

The choice of the institutional concept as the methodological base of the study and our model of analysis determines the overall structure of the research. It incorporates three major themes and their respective key elements (as pointed out below) which are addressed in separate clusters of chapters of this research.

1. CIVIL SERVICE REFORM CAPACITY

- Institutional framework – existing formal and informal institutions (combination/set of rules, norms, traditions, values and behavioral patterns) and historical legacy;
- Collective or individual human action - willingness and readiness of the key agents – policy makers and stakeholders to undertake action in the desired direction;

- Resources available - funds and expertise.

Thus, in compliance with the institutional approach and respectively our model of analysis, we intend to devote priority attention to the study of formal and informal rules, norms, traditions, values, culture, behavioral patterns, historical legacy prevailing in state administrations of each of the countries concerned, complemented by the analysis of agency (leadership) related factors and available resources. Exploration of these factors will allow us make certain assessment of the capacity for civil service reform in the Russian Federation and the Czech Republic.

Respectively country specific chapters will comprise sections dealing in particular with historical development of the nationhood and national state administrations; existing constitutional and political framework; economic situation; legal status of the civil service; political-administrative relations; societal and bureaucratic traditions, values and culture; civil servants' management and incentive mechanisms; ethical standards and accountability of state officials; their relations with civil society, etc.

2. CIVIL SERVICE REFORM IMPLEMENTATION

- Who or what initiates reform and why - reform's major stakeholders; dominant administrative tradition (governance or management); category of reform (budgetary/financial, structural, procedural/technical, relational);
- Dimensions (contents) of reform - objectives of reform and outputs expected; methods of achieving planned objectives; reform management and pace of reform; results achieved up to date;
- Possible external influence - borrowing of specific concept; borrowing of broad direction approach.

In line with the guidelines of the Protocol for comparative studies and our methodology, chapters dealing with the issues of civil service reform implementation in Russia and the Czech Republic will incorporate analysis of such issues as the reforms' driving forces and stakeholders (both internal and external); reforms' dimensions and management; processes and mechanisms of reform advance, encountered problems and results achieved so far.

3. CIVIL SERVICE REFORM COMPARISON

- Similarities and differences - in reform capacities, contents, preparation and implementation, achieved results (including comparison of key independent variables);
- Possible explanatory factors (of differences and similarities);
- Concluding remarks (findings correlated with hypothesis and research sub-questions).

In order to reach the research objectives (in particular with regard to finding answers to the research main question and sub-questions) we will need to carry out a comparison of the reform capacities and dimensions, processes of reform preparation and implementation in the two selected countries and their current outcomes. Therefore, two separate chapters will be devoted to identification of major similarities and differences in reform capacities and reform implementation (including their dynamics and achieved results) respectively. Matching impact of key variables on the process of civil service reform and administrative change in each of the two countries will lead us to certain explanatory factors and subsequently permit to substantiate major findings and conclusions.

With regard to selected variables we may at this point assert that they may be directly or indirectly linked to factors including: the character and essence of the political regime and development of political institutions; resoluteness of political leadership either to be a driving force of changes or support other stakeholders; type of relations between the policy makers and top civil servants; extent of domination of interest groups; prevailing behavioral patterns and informal norms and rules; maturity of civil society; extent of internal and external pressure; and available resources (funds, expertise and others).

We hold the view that in cases when the existing administrative culture remains hostile to the goals and spirit of reform and there are no committed reform stakeholders (or they are few and weak) then reform implementation tends to be formal and minimalist with no real changes to be achieved in practice. This we anticipate in differing degrees in both Russia and the Czech Republic. We also reckon that political leadership as well as internal and external pressure often exert a decisive impact on the pace of advance and results of civil service reform.

2.3 Civil Service Reform: Definition, Dimensions, Problems

2.3.1 Contemporary Views on Civil Service Reform

The civil service system is one of the core concepts of public administration science and as such has been drawing increased attention of the scholarly community in recent years. Our knowledge of civil service systems is still very insufficient and fragmented. Civil services have attracted the interest of comparative research scholars especially in the 1960s. However, this research was soon largely abandoned mainly because of failure to develop a convincing theory and subsequent period of "crisis of confidence". There has been little follow-up to the early comparativists' efforts since then.

For the purpose of this research it would be appropriate to distinguish between civil service as an "institution" and as an "organisation" (originally proposed by the 1993 Economics Nobel prize winner Douglas North). As an "institution" it means "a set of formal and informal rules", as a public organisation it "emerges to administer and enforce these rules". This distinction is essential for understanding the problems of the institutionalisation and transformation of civil service systems in CEE countries.

Our understanding of civil service "membership" (cadres, personnel) encompasses career, appointed functionaries at both central and sub-national levels of government and excludes elected, political officeholders. Military personnel, except civilians working for the armed forces, are not part of the civil service either.

In dealing with issues of civil service reform we proceed from the definition of a civil service system as "mediating institution that mobilizes human resources in the service of the affairs of the state" given in the publication "Civil Service Systems in Comparative Perspective" edited by Bekke, Perry and Toonen.

"In many ways, civil service systems are shaped and stabilized by formal or informal norms and values, historically developed and adapted to particular societal experiences and beliefs" (Bekke,1999: 4). Offe argues that in those cases where institutional norms, embedded in specific organizational arrangements, deviate permanently and substantially from those held by people acting within an institution, institutionalization is precarious (Offe, 1996: 212). This means that although new institutions may be in place, they do not necessarily always perform the role assigned to them from the very beginning and sometimes have a counter productive effect.

In reviewing modern theories and views on civil service reform we can not bypass two leading publications on the subject – *"Civil Service Systems in Comparative Perspective"* edited by Bekke, Perry and Toonen, and *"Public Management Reform. A Comparative Analysis" by* Pollitt and Bouckaert. Both publications explore a set of key issues and themes related to public management reforms and make a fair contribution to further elaboration of methodological tools for comparing administrative systems.

In studying various aspects of civil service reform we start with the definition of public management reform as "deliberate changes to the structures and processes of public sector organizations with the objective of getting them to perform better" (Pollitt & Bouckaert, 2004:16).

It is generally recognized that the last two decades (at least) have seen waves of administrative and civil service reforms spreading across many industrialized and less industrialized nations in the world. Many prominent scholars in the field have argued that constraints, contradictions, limitations are already in-built in practically any public administration reform initiative even before the reform actually starts. "It would be inadequate to conceive of reforms as simply a string of connected actions. If they are to be understood then they must also be considered as processes of debate to which different participants may bring, first, different objectives (including achievement of symbolic purposes) and, second, different frameworks and standards for identifying and accepting relevant 'evidence'. (Pollitt & Bouckaert, 2004:16).

In their study Pollitt and Bouckaert propose a model of reform meant to show a complex of factors (variables) that may influence (drive or constrain) public management reform and change. These factors are grouped into clusters depending on their relation to:

- *socio-economic and demographic forces;*
- *political and intellectual trends;*
- *elite decision making;*
- *chance events;*
- *administrative factors.*

Pollitt and Bouckaert deliberately place the *process of elite decision making* at the centre of their model of public management reform. In their view "most of the changes we are concerned with have been predominantly 'top down' in the sense of having been conceived and executed by executive politicians and/or senior civil servants" and therefore being "essentially voluntaristic activities". At the same time the authors admit that the elites "may be heavily influenced by ideas and pressures from elsewhere and, furthermore, their plans may be blown off" (Pollitt & Bouckaert, 2004:26).

Some comments made by the authors deserve special attention. First, "it is the exception rather than the rule for reform schemes to be comprehensive, even in intent". Second, "it is easy to exaggerate the degree of intentionality in many reforms" (ibid). The second point is explained by Goodin who states "Institutions are often the product of intentional activities gone wrong – unintended by-products, the products of various intentional actions cutting across one another, misdirected intentions or just plain mistakes" (Goodin, 1996:28). Another note worth observation in connection with study of reform is that "remedies imposed for perceived problems often generate new, and opposite problems" (Hesse, Hood, Peters, 2000:14).

Some reforms are top down, some are bottom up. Some are designed to enhance state legitimacy, while others attempt to raise administrative efficiency. Some are cost cutting exercises, others are aimed at raising the quality of public services. With regard to various models of civil service reforms Pollitt and Bouckaert point to the limitations of discussing reform exclusively in terms of "*more or less NPM* " ("new public management" – A.K.) ….there are alternative and positive concepts of modernization, one of which we describe as the "Neo-Weberian State" (Pollitt & Bouckaert, 2004:3).

The same authors emphasize that "we need to pay attention to the legitimation aspects of management reforms. Simply announcing, discussing and beginning to implement reforms may bring benefits to some politicians and public servants, even if the later, more 'substantive' effects are elusive or counterproductive" (Pollitt & Bouckaert, 2004: 25). Speaking about potential implications of management reforms and causes of resistance they note "well-oiled networks of information and influence are disturbed, and new, less certain ones put in their place. Politicians who were used to one configuration of authority within the state agency that most interested them now have to get used to a new pattern, and possibly one that will be more difficult for them to influence or communicate through." (Pollitt & Bouckaert, 2004:34).

Reformist minded politicians need to make allowances for the strength of the forces of tradition, inertia and habit. In connection with this it is necessary to bear in mind that reform related rhetoric, decisions and actions frequently diverge. Proponents of the 'path dependency' theory proceed from the assumption that heavy investments made over decades in the existing institutions, rules and the ways of doing things usually create a huge disincentive for change (although there may be also opposite cases, see Raadschelders, 2007). Still there exist countries where reform was launched and implemented reaching if not all but at least some of the main objectives. At the same time there are countries where, in spite of intensive rhetoric, reforms have virtually changed little or even made things worse.

Finally we would like to underline once again that civil service reform is widely recognized to be a highly political issue, which affects the balance of power between various actors. "What has become more and more obvious is the extreme importance of variation among political regimes as a major explanatory factor for variation among public bureaucracies…" (Heady, 2001:444). Successful reforms require both political commitment at the highest level and the capacity to concentrate and manage resources in order to implement and sustain reform. Civil service reform is a complex process that does not easily achieve its objectives by specific deadlines.

2.3.2 Civil Service Reform Dimensions and Outcomes

Universally acknowledged "standards" with regard to the dimensions and basic elements of civil service system and civil service reform do not appear to exist. As stated earlier an appropriate definition of a civil service system is a "...mediating institution that mobilizes human resources in the service of the affairs of the state in a given territory" (Bekke, Perry, Toonen, eds. 1996:2). In the opinion of the authors of this definition the essence of a civil service system is determined by social and legal rules (i.e. prescribed behaviour as well as explicit and implicit norms) which tend to persist.

Other approaches to defining dimensions of a civil service system and civil service reform comprise in particular:

- People, structures, behaviour, power (*Peters*);
- Relation to the political regime, focus for personnel management and qualification requirements *(Heady);*
- Relationship with policy-makers, internal organisational dynamics, and relationship with civil society (*Pierre*);
- Rules, structures, roles and norms (*Morgan, Perry);*
- External pressure, internal dissatisfaction, reform strategy, mechanism for managing reform, feedback and evaluation (*Les Metcalfe*);
- Interests, actors, resources and "administrative" traditions (*Goetz*).

Civil service reform is an organic process, which can not be studied in isolation from its complex environment. At the same time civil service reforms are subject to the process of bargaining where the essence, clarity and scale of proposals as well as their objectives may be frequently lost or replaced.

Often civil service reforms are part of broader development programmes (in particular, administrative reforms) and are closely related and interdependent with judicial, municipal, budgetary, regulatory reforms, re-organization of central government and others.

A number of important observations are worth mentioning in this respect.

Pollitt and Bouckaert highlight the complexity of the reform implementation process and point to numerous cases (becoming rather a rule than exception) of departures from the original plans of reform in course of their execution. Even when reform itself and its implementation are appropriate it may contradict or detract from other reforms that are being carried simultaneously (Pollitt & Bouckaert, 2004:36).

In the opinion of Toonen, the largest impact on variation in reforms is made by the following two separate dimensions:

a) Substance: different subject matter of reform (managerial reform; policy reform; institutional reform; regime reform) and

b) Process: different modes or approaches to public sector reform (comprehensive, functionalist, gradualist reforms).

He further argues that the following factors are most relevant in explaining variation in public sector reform:

1. External pressures e.g. economic reform and European integration;
2. Political choice e.g. ideology and leadership;
3. Internal dynamics e.g. reform logic and 'path dependencies';
4. Institutional opportunities and constraints e.g. types of administrative systems;
5. Historical context e.g. historical legacies and traditions (Toonen, 2004).

We share these arguments, which derive from the institutional approach to the research of public sector reform. It is noteworthy that the structure of our study and the selected format for comparison (see section 2.2.7) largely reflect the existence of the above factors (singled out by Toonen). Similarly, in our study we define two basic characteristics of a civil service reform which deserve special attention – the extent (scope and dynamics) of reform and the character (dimensions) of reform.

Concerning the eventual outcomes of public management reforms Pollitt and Bouckaert note that "in practice, the 'final results' of the reforms are frequently difficult to identify with any confidence" (Pollitt and Bouckaert, 2004:37).

As many observers and experts indicate when expectations regarding reform goals and results are high and promises are far reaching, then in evaluating or assessing reform implementation it is very difficult to state anything else but good results. As far as information concerning reform results is concerned Pollitt and Bouckaert distinguish four levels of results ((Pollitt & Bouckaert, 2004:105):

a) operational (more outputs are obtained for the same inputs);
b) improved processes of management or decision making or service provision (streamlining, reducing transaction costs, making more convenient etc);
c) change in the overall capacity of the political or administrative system (e.g. redesigning a pattern of the institutions, higher potential to respond to and resolve problems etc);
d) degree of advance of the system towards some ideal or desired state (criteria of that level are rather ideological or doctrinal).

The last two levels of results are rather abstract and include also value shifts and cultural transformations.

Some scholars and experts divide improvements in the system into two groups: structural improvements and cultural change. The former means various large scale reorganizations whereas the latter denotes such things as shift in civil servants' values, changes in the accepted or prevailing standards of behaviour and attitudes, etc.

2.3.3 Peculiarities of Post-Communist Administrative Development

The Soviet system has been regarded by many experts to have the most "advanced" model of untamed bureaucracy. For example, Hough viewed the Soviet Union, "as a total bureaucratic system in which the leaders themselves rose through the bureaucracy and are part of it rather than a parliamentary system in which a cabinet of generalist political leaders gives direction to a pliable bureaucracy" (Hough, 1973: 143).

Needless to say other countries of the Soviet block developed similar bureaucratic systems, in some cases even more rigid and less tolerant than the Soviet "model" proper.

In spite of profound changes in the political and economic systems of CEE countries in the 1990s the heritage of the bureaucratic past could not have been easily got rid of and repeatedly manifested itself in particular in the continued domination of old style informal rules and traditions in the administrative process. These rules and traditions largely defined the strength or weakness of major political actors and perspectives of success or failure of reform. Connected with the past was also the phenomenon of persistence of old elites and high level of politicization of the state bureaucracy in almost all CEE countries;

The "communist type" patterns of governance appear to have reproduced themselves to various degrees in the changed context of the region. In connection with this Toonen remarks "Given the magnitude of the changes and transformations at the end of the last decade, the degree of continuity and influence of the past comes as somewhat of a surprise." (Toonen, 1993:165).

The establishment of new public and civil service institutions in the former communist countries occurred in the overall setting characterized in most cases by high degree of political, economic and social turbulence, increased personalization of politics, prevalence of uncertainty and short time horizons. The continuous instability and fluid situation in many areas of public life in these countries, as well as deeply rooted traditions of "selfish" bureaucratic behaviour, patronage and neglect of the interests of citizens served as distinct factors affecting the overall environment of reform.

It appears that the bureaucracy in the majority of post-communist countries has been one of the major actors of the political and economic transformation of the 1990s. Besides, it largely succeeded in transforming the administrative power into economic power through the semi- or quasi- privatization.

Preparation and actual accession of many CEE countries to the European Union have been, beyond any doubt, an important factor affecting development of state administration in these countries. However, the assessment of practical impact of accession on the modernization of CEE states' bureaucracies remains controversial. Some authors (e.g. Goetz, Meyer-Sahling) reckon that the impact of EU accession may have heightened rather than lessened incongruities and asymmetries in the development and functioning of public service institutions in CEE countries.

Throughout the 1990s the civil society in the majority of CEE countries was still in the process of formation and in most cases turned out to be too weak to act as a powerful independent policy making factor. In connection with this Pierre remarked, in our view correctly, that "the best guarantee for an improved public service is not internal administrative reforms but a stronger and more vibrant civil society that can hold public officials accountable" (Pierre 2001: 181). When such mechanism of accountability is absent then political actors in post-communist countries quite often either "deliberately fail to design institutions: thus building in enormous license and opportunity to seek rents" or "deliberately set up institutions with very limited enforcement powers that exist mainly on paper "(Grzymala-Busse, 2004: 3).

Another noteworthy consideration concerns the role of budgetary constraints and pressure in stimulating administrative reforms in CEE countries, which appears to differ substantially from the experience of other industrially developed countries. There exists sufficient evidence for the claims that instead of encouraging reform and modernization of the state bureaucracy, severe budget deficit in the post-communist countries served as an additional obstacle complicating the start of reforms. "Economic pressure could at first glance be a catalyst for the initiation of administrative reform processes, as there still are serious budgetary pressures in all candidate states. ... However, even though economic conditions at first glance constitute a potential incentive to politicians to move administrative development forward, the specific nature of the administrative reform processes in Central and Eastern Europe has so far created the reverse effect." (Verheyen, 2001:40).

The same author continues - "It is important to note that building a professional, stable and impartial administration is costly and requires a considerable investment, even if one opts for the creation of a limited 'career' civil service. The costs of recruitment, training and the employment conditions associated with a career civil service are considerable, both in the immediate term and for the future. The notion of administrative reform as used in Western Europe and other OECD states, the main objective of which is usually saving money, therefore does not apply to Central and Eastern European states. The economic conditions in these states thus make the implementation of reform strategies more difficult, rather than easier" (ibid).

There exist basically two types or groups of issues that should be resolved in the transformation of state bureaucracies of CEE countries.

One group is more technical – formation of a modern civil service as such, adoption of respective legal framework, introduction of principles of merit, enhancement of performance and efficiency of state bureaucracy, application of client oriented approaches, introduction of new IT technologies, etc. The other relates to more fundamental problems of the place, role, responsibility and relationship of the state bureaucracy to the society and politicians, as well as of prevailing within the bureaucracy informal rules, values and ethical standards. The second group in our opinion has greater significance in the case of post-communist countries as far as profound and sustainable reform of their civil services is concerned. Touching upon this issue Hesse writes "The challenge with which public institutions are faced in Central and Eastern Europe, is to redefine their role in society, or, more concretely, their relationships with politics, the economy and the civil community..." (Hesse,1995: 5).

2.3.4 Assertions and Hypotheses

Our initial analysis enables to single out a number of assertions, which characterize developments in the post-communist countries, and in particular in Russia and the Czech Republic, related to the evolution and reform of their state administrative systems. These assertions and hypotheses concern the main question and sub-questions of the research and are linked to its major topical areas - capacity, implementation, comparison.

CAPACITY

We consider that the capacity for civil service reform in both the Russian Federation and the Czech Republic was overall low throughout the 1990s. We assume that this capacity may have been weakened by:

- persistence of outdated norms, traditions, behavioral patterns and relationships;
- shortage of resources both financial and human;
- lack of political will to launch reform;
- deficit of committed stakeholders;
- immature civil society.

IMPLEMENTATION

We may expect that implementation of civil service reforms in Russia and the Czech Republic:

- will proceed along the directions and will set the objectives largely borrowed from the experience of the advanced industrial nations;
- will meet notable resistance of the large part of state bureaucracy and "vested" interests;
- will be more centered on the reform process than on achieving reform results;
- will focus on less radical and more "convenient" elements of the reform agenda.

COMPARISON

We assume that:

- there are more similarities than differences in the way reforms have been prepared and implemented in Russia and the Czech Republic;
- a set of key variables affecting the process of reform preparation and implementation in both countries is to a large degree identical, though their actual impact may have differed;
- reform implementation in both countries may have produced certain side effects and unforeseen results;
- extent and intensity of reform implementation largely depend on the continuation of political support to the reform at the highest possible level;

We proceed from the assumption that many informal rules and networks (widespread in the bureaucracies of the former communist countries) survived political upheavals and remain the base of relations across state administrative institutions in the Russian Federation and the Czech Republic. These networks may be "reinforced" by a high degree of politicization of the Russian and the Czech civil services. With all other factors being equal, the role of leadership will be crucial for the success of civil service reform. Therefore, preferences and personal qualities of politicians (who act under specific conditions and whose discretional power is also limited) do matter.

We acknowledge the existence of considerable and often essential differences between the two countries. Russia is a federal state and presidential republic with huge territory, multinational multi-religion ethos, huge reserves of oil, gas and other natural resources, state budget proficit in recent years. In comparison the Czech Republic is a parliamentary unitary state with relatively small territory and rather homogeneous population, budget deficit and scarcity of natural wealth (to mention just a few features). In spite of these differences we consider that the contents, pace of preparation, advance and initial results of civil service reforms in the two countries will have a lot of similarities.

We assume that the similarities in reform preparation and implementation can be in particular attributed, firstly to the largely common roots of administrative culture and mentality traceable to the heritage of monarchist (Austro-Hungarian and Russian Empires) and recent communist regimes, and secondly to peculiarities of the political setting in these countries manifested above all in seriously constrained opportunity for competition between various political parties and groupings.

Distinctions in reform preparation, advance and outcomes may result from such factors as external influence, differences in the constitutional order and political regime, as well as resources available.

2.3.5 Defining Principal Terms and Concepts

A clear definition of major/key concepts and terms that are widely used in this research is a precondition of a sound methodological approach. Most frequently used terms and notions comprise:

Bureaucracy – one of the meanings of this notion most widely used here is "a corps of civil servants occupying top and mid-level management positions";

Bureaucratic accountability – is commonly understood as the answerability of civil servants to the "public" for their actions or inactions, for which they are subject to external or internal sanctions;

Central government – central governing and administrative authority at the level of the nation-state (usually the highest or one of the highest levels of the executive branch of power);

Civil society – "a wide array of non-governmental and not-for-profit organizations that have a presence in public life, expressing the interests and values of their members or others, based on ethical, cultural, political, scientific, religious or philanthropic considerations. Civil society organizations, therefore, refer to a wide of array of organizations: community groups, non-governmental organizations (NGOs), labor unions, indigenous groups, charitable organizations, faith-based organizations, professional associations, and foundations" (World Bank, 2005);

Civil service – may be understood as "an institution of governance programmed by rules which are, by definition, reasonably stable over time and take on the character of roles, rules, norms and expectations about civil service attitudes, behaviors and functions in discharging state business" (Bekke et al., ed, 1996:49).

Three connotations of the term 'civil service' can be defined in our context:

- social and legal state institution;
- professional activities, aimed at exercising the authority of state organs of power;
- organisation with relevant structure, personnel and allocated resources.

Note: It is important to distinguish between "higher civil service" - a relatively small group of top bureaucrats and the 'rank and file' civil servants;

Civil service reform – "deliberate changes to the structures and processes of public sector organizations (in our case civil service – A.K.) with the objective of getting them (in some sense) to perform better" (Pollitt & Bouckaert, 2004:16);

Civil service system – "mediating institutions that mobilize human resources in the service of the affairs of the state in a given territory" (Bekke et al, ed, 1996:2);

Conflict of interest – a situation in which professional decisions or actions risk being influenced by personal interest or considerations;

Context (environment) - in comparative analysis generally refers to all external influences that affect processes and management, such as societal values, norms, religion, political culture and economy;

Deconcentration – delegation of the execution of state administration from the center to vertically lower tiers of state administration;

Decentralization – delegation of the execution of public administration from state administration bodies to local government;

Development - is viewed as an ability of a system to grow or adjust to new demands put upon it, or "an ability for movement from a less desirable to a more desirable general situation in the political system" (Heady, 2001:106);

Good governance - process or system of governing, which ensures that an organisation promotes generally recognized positive values such a legitimacy, rule of law, fairness, protection of human rights, efficiency, effectiveness and conduct excluding conflict of interest;

Institution – a social organisation with an inherent set of formal and informal rules. Also "persistent and connected sets of rules (formal and informal) that prescribe behavioural roles, constrain activity and shape expectations." (Keohane,1989:163);

Path dependence – a concept emphasizing the connectedness of the present and future with the past; in particular, that the past limits the range of choices, which are available at present;

Political regime – a system of practical exercise of political authority, which determines real balance of power between the executive, legislature and judiciary and extent of accountability of the state to citizen/civil society;

Public administration

a) the activities of the government and public servants ;
b) the totality of institutions, respective personnel, necessary resources and adopted procedures at all levels of government ;
c) the study of a) and b).

Public administration reform - reform of the 'public administration' perceived mainly in the second meaning (see above);

Stakeholders – individuals or organizations with a distinct interest or concern in the outcome of a particular process;

2.4 Conclusions

The methodology applied in this research is largely based on the so-called institutional analysis, which implies "...the notion of systems ruled by norms, culture and traditions rather than by organizational goals and rational perspectives" (Bekke 1999: 4), and hence attaches high importance to the environment, in which any institution, civil service inclusive, is embedded.

More specifically, institutional factors are seen as the main source of explanation for the peculiarities of political and social processes in general and of administrative reforms in particular. Institutions are viewed from a broad perspective so that they are considered to incorporate apart from organisations and structures also norms, rules, values, traditions and ideas. Furthermore, the institutional approach emphasizes persistence of rules, values and behavioral patterns over time. In other words, it assigns an important role to the concept of *path dependency*. The meaning of the latter can be condensed to a phrase "past choices limit future options".

Civil service reform is a complex process, which cannot be studied in isolation from its environment. Therefore, particular attention in our research is paid to the retrospective study of the existing constitutional and political framework, economic situation, administrative system, culture and mentality, traditions and behavioral patterns which make up the actual setting of each of the civil services. The analysis of the relationship between an administrative system and its context is particularly important for identification of key factors/variables affecting the success or failure of reform.

At the same time the particular path of civil service reform is usually a result of complex interactions of various actors and factors within certain boundaries imposed by the existing institutional setting and "path dependency". Although we are far from accepting human action as the main determining factor, our approach (in compliance with neo-institutional concept) overcomes opposition of institutional factors and individual action. The latter will be given proper attention wherever appropriate.

The study draws widely upon the theory, methods and experience of comparative administration accumulated in the publications of most prominent scholars and experts in the field. In particular, the research widely applies inductive method that relies heavily on descriptive data about actual situations and behaviour. It presupposes advance from specific observation to identify differences and commonalities and then moving to explanations and interpretations. The process usually concludes with generalizations explaining relationships amongst units observed.

In brief, our methodological and analytical framework implies that study of existing rules, norms, values, traditions and authority relationships which exert significant influence on formation of administrative systems and civil servants' behaviour and that also persist in time is a necessary precondition for proper understanding of civil service systems and their transformation. It also implies that various institutions define and limit choices available for policy-makers but they do not deny the role of individual and collective human action.

With regard to peculiarities of the development of administrative institutions in the CEE region it is necessary to take into account that in the 1990s this development in the absolute majority of post-communist countries was affected by instability and a fluid situation in many areas of public life, serious economic difficulties, absence of mature civil society, increased personalization of politics, prevalence of uncertainty and short time horizons. By the mid 1990s state bureaucracy in many CEE countries appeared to have been least of all affected in real terms by the transformation as compared to other public institutions and sectors of the economy.

Two more remarks are noteworthy. First, preparation and actual accession to EU have influenced considerably administrative development in many countries of the region. Second, contrary to the experience of other industrially developed nations severe budgetary constrains in CEE states have served as an additional obstacle (not as a stimulus) in the way of initiation of civil service reforms.

CHAPTER III
THE RUSSIAN CIVIL SERVICE: EVOLUTION, STATUS AND REFORM CAPACITY

3.1 Introduction

This chapter incorporates an analysis of the Russian civil service system and its capacity for reform. The methodology applied herein has been presented and explained in the previous chapter.

Particular attention in the chapter is paid to the study of institutional environment and historical development of the Russian state administration. We proceed from the premise that "historical and contemporary comparison must always take into account the context in which specific phenomena occur. Studying development over time, juridical, political, economic, social, cultural and even geographical conditions must be part of the analysis. If we want to compare specific administrative arrangements in time and place, it is only by reference to these contexts that we can appreciate the contingency of events" (Raadschelders & Rutgers, 1996: 92).

First, we trace the development of the Russian state and its administration since the formation of the Moscow princedom and the unified Russian state in the 10-14th centuries. Then, we study peculiarities of the constitutional order and political system in Russia, economic development in the 1990s, including repercussions of privatization. Afterwards, evolution of the Russian civil service during the post-communist period is analyzed with particular accent on the issues of legal framework, political-administrative relations, HR management, education and training.

Further, we examine the existing administrative traditions and culture, prevailing in civil service ethical standards and the extent of accountability of civil servants. This is followed by an overview of the relations between civil service and civil society and estimation of resources available for implementing civil service reform. The analysis culminates with conclusions regarding key problems and constraints in the development of the Russian civil service and an overall assessment of its reform capacity.

Most of the information and data provided in this chapter covers the period from the early 1990s till mid 2000s, whereas overview of the recent changes in the Russian civil service is given in the following chapter.

3.2 Evolution of the Russian State and its Administration

The Russian history can be presented in a way as *a succession of periods of enforced (as a rule top down) modernizations and periods of stagnation or decay in social, economic and political spheres.* A number of factors significantly influenced this evolution: the existence of huge, largely unpopulated territory inhabited by multinational multi-religion ethnos; permanent external threats and frequent internal conflicts; and the nature of the relationships within communes that were the basic element of social organization in Russia in the times of monarchy.

The state bureaucracy has been one of the leading actors in Russian history and played a decisive role particularly during its crucial moments including the periods of sweeping reforms. The process of the evolution of the Russian bureaucracy deserves special attention because it is precisely here that we can find the origins of many of its present day features and peculiarities.[1]

The Kievan Rus

Civil servants in their very primitive form appeared on the territory of Russia as early as the 9-10th centuries. Around that time the southern and northern princedoms of the Slavs, situated on a huge territory from the Baltic Sea to the Black Sea, were united by Prince Oleg, who in A.D.880 conquered Kiev and made the city his capital. At that time the Russian state (*Kievan Rus*) consisted of principalities loyal to Prince Oleg and to his successors.

With the appearance of a unified Russian state the need for the Sovereign and his closest circles to have a group of subordinates, who would assist them in running the state, became more than evident. Therefore, apart from servants who performed purely personal (hunters, cooks, housekeepers, etc.) and military functions, there gradually emerged servants responsible for broader tasks - tributes registration, arbitration of disputes, maintaining relations with neighbouring countries - although these servants were still considered personal servants of a ruler.

In 988, Grand Duke Vladimir introduced Christianity to Russia. An attempt to create a comprehensive system of laws was made by Princess Olga (10th century). In executing governance Russian Princes discussed most important decisions at the meetings with their retainers (*boyars*), prominent warriors and most respected and wealthy citizens. Later these meetings led to a special advisory council (*duma*). Rivalry between the principalities never ceased and in the 12th century the country disintegrated. From that time on the Moscow principality grew in size and significance and gradually started to play a central role among all other principalities.

Introduction of Christianity contributed to further legitimization of Russian Dukes (Sovereigns), who in compliance with Byzantine tradition were presented as Holy nominees empowered by the Lord not only to organise defence against external threats, but also manage internal affairs and maintain public order.

The Tatar-Mongol Domination and Strengthening of the Moscow Princedom

The invasion by the Tatar-Mongols in the 13th century had a great impact on the development of the principalities and the whole Russian state. *The subsequent rule of the Golden Horde cut Russia off from the outside world for more than two centuries.*

It is interesting to note, that two northern princedoms - Novgorod and Pskov managed to preserve their independence in the face of the Mongols' invasion. They established virtually a republican state of governance. Novgorod princes were elected

---

[1]Sources of information: www.russia.net/history, www.pts.mipt.ru, www.studyrussian.com/history.html

by citizens for a fixed period of time. The power of a prince was restricted by special agreements. All issues of great importance were solved at gatherings (*vyeche*) of citizens. That was in fact the first attempt to introduce elements of legitimacy of governance in Russia.

In 1380 by the decisive battle on the river Don, Prince Dimity defeated the Mongols. However, it was not until 1480 that Moscow was strong enough to throw off the Mongol's yoke. Russia became again a centralised state in late 15th century under the rule of Ivan III. During his reign *the boyars* (landed aristocracy) retained their overall influential role in the management of state affairs. Seniority among *the boyars* depended on the family history of service to the court and on landownership. Ivan III also opened membership of this elite group to anyone who held a high administrative post. In 1547 his grandson Ivan IV (*The Terrible*) was crowned the first tsar of all Russia and his power became absolute.

The Reign of Ivan the Terrible

The establishment of state administration in Russia as a distinct structure acting in line with certain norms and rules occurred during the 16th century in the reigns of Ivan III, Vasily III and Ivan IV. '*Prikazy*', special organs of governance and control were established and organised along functional and territorial lines. There were *prikazy* responsible for search of criminals, for relations with other countries, for state finances, for mail and transportation, etc., as well as those in charge of different regions of Russia. New entrants to *prikazy* were recruited basically from two sources; either from princes' ex-warriors and personal servants or from the families of noble landlords (*boyars*).

Soon a true bureaucratic style developed in the *prikazy* with rigid vertical subordination and strict observance of instructions and orders in horizontal communications. A general rule was followed, such that the more noble a man was, the higher post he could get in the administrative structure. Regional and local reform at that time introduced elements of self-governance by citizens (*zemskiye and gubskiye izby*).

In his struggle against the influence of the feudal aristocracy (*boyars*), Ivan the Terrible confiscated much of their property and land, granting it to those who served him. These beneficiaries gradually formed a new social group or nobility (*dvoryanstvo*). In his late period of rule, Ivan IV established a regime of extreme terror, which brought the country to a state of almost total political and economic ruin.

The Rule of Peter the Great

After the so-called 'time of troubles' (most of the 17th century) Peter I (the Great), one of Russia's most enlightened and controversial rulers, assumed the throne in 1696. As a young man he had spent two years in Western Europe (England, Holland, Sweden, Prussia) learning, among other things, state administration practices. Subsequently he became determined to carry out sweeping reforms in Russia and to extricate the country from its isolation.

Unlike his predecessors, Peter the Great admired and respected the ideas and culture of Western Europe. *The first reforms of the Russian state administration occurred during his reign.* Peter the Great adopted a concept of the enlightened despot who served the state. During his rule boyars lost their privileged position.

Aristocratic origin was no longer taken into account when considering appointments to high level posts in state administration. Instead length of service, qualification, performance (for the first time) and loyalty to the sovereign counted.

In 1722 Peter I approved a Table of Ranks which introduced 14 ranks of state service and spelled out in detail its order and conditions. This was the first comprehensive legal document that created a sound basis for the development of the professional state administration in Russia. In preparing this document his experience of other countries, e.g. England, Denmark, Prussia, Sweden, the Netherlands was carefully analysed. The Table of Ranks introduced clear division between military and civil (public) service and imposed compulsory lifelong public service on the nobility. From that time onwards service to the monarch required appropriate education and provided for, although on a limited scale, ennoblement of commoners. During this period the state apparatus became more professional and specialised, but at the same time less flexible and more oriented towards meeting its own corporate interests. The hierarchical structure of governance, totally subordinated to the Emperor (Tsar), included the Senate (the highest organ of central governance), Collegia (displaced prikazy) and Governors in provinces (*gubernyi*). Peter the Great defined the distribution of authority, responsibility and functions within the administrative apparatus and forced public servants to work in accordance with state laws, standard norms and rules. It is interesting that according to a decree signed by Peter I on 17 December 1722, violation of state laws by a civil servant would immediately entail death sentence "regardless of his/her merit and record of service" (Collection of Laws of the Russian Empire, 1830).

The Period of Enlightenment

Two important processes dominated in Russia in the 18th century - territorial expansion southward, eastward and westward and cultural westernization of the Russian elites. This period is known as the period of 'enlightened absolutism'. Catherine II (the Great) (1762-1796) tried in particular to promote the formation of a civil society in Russia. Though many of her ideas fell short of changing the existing practice, she did lay the foundations for subsequent reforms in the next century.

Catherine II continued Peter the Great's reforms of the Russian state, further increasing central control over the provinces and contributing to the resurgence of the Russian *nobility*. The latter received expanded rights and authority, were no longer required to serve the central government, but took over most government functions in the provinces.

Different social groups at that time attempted to introduce constitutionalism, to abolish favouritism and to enforce respect to the rule of law. This largely failed because of divisions among the nobility, its dependence on service to the ruler and fear of power being delegated to impersonal institutions. One of the remarkable developments concerning the state bureaucracy at that time (mid 1760's) was introduction of *fixed permanent remuneration to civil servants of all categories and granting the right for pension after 35 years of service*. In accordance with the decree of Catherine II (1763), in performing their duties civil servants should be guided by "needs and benefits of the state".

In spite of these improvements the majority of the Russian civil servants as in many other countries were forced to extract rent from the exercise of their duties, since the government paid them fees below even subsistence level.

Developments in the 19th Century

In the first half of the nineteenth century the ideas of individual and political freedom and social equality were penetrating Russia from Western Europe. Yielding to the expectations and demands of the society, Emperors Alexander I (1801-1825) and Alexander II (1855-1881) embarked upon far-reaching reforms. The architecture of the modern Russian state was largely drawn in the early 19th century, when Alexander I established in 1802 the first eight ministries, which incorporated real administrative power. This move was followed by the establishment of the State Council analogous to western parliaments (although the analogy was limited as its members were appointed by the Emperor who also approved its decisions). Later came judicial reform (the introduction of jury trials, independent judges and a professional class of lawyers) and the emergence of local self-government – '*zemstvo*' - exempting local affairs from a considerable amount of central government control. In 1861, Alexander II abolished serfdom, which was a major obstacle to industrial revolution in Russia.

Nonetheless, resistance to the reforms was enormous, particularly from the nobility which made up the core of the state apparatus. It took half a century to prepare and adopt a law on abolition of serfdom. The last two decades of the century were marked by virtual retreat from the progressive changes. Russia entered the 20th century with no legitimate government in place and performance of the state administration being considerably below expectations of the society.

Under the monarchy the role of state administration was constantly growing. The first half of the 19th century witnessed a large expansion of the corps of state administrators, the absolute majority of which represented nobility. By mid 19th century the top state bureaucrats made up the largest part of the ruler's closest circle and together with the Monarch were determining the internal and external policies of the country. The state bureaucracy enjoyed a privileged status and was actually exempt from effective social control. Among most typical features of the bureaucracy were strong authoritarian tendencies, neglect of the needs of common people, opposition to civil society and virtual absence of any ethical standards of conduct.

The conservative tendencies in the tsarist bureaucracy were sustained by its privileged position, absence of oversight by any representative organ, atmosphere of despotism, total obedience to the superior and suppression of independent thinking.

Nicolas II and October Revolution

The number of Russian civil servants in the 19th century increased seven times and amounted to about 385 000 by the year 1900 (Zaionchkovsky, 1978:221). In the early 20th century 14% of the state budget was allocated to finance the state administrative apparatus, compared with 3% in England, 5% in France and 7% in Italy and Germany (Zyrianov, 1994:147).

In 1891, Nicholas II acceded to the throne. His government became increasingly unpopular because of opposition to any reform and because of the defeat of the Russian fleet in the war with Japan. General discontent and labour unrest culminated in the revolution of 1905 and forced Nicholas II to grant a constitution. Thereby Russia acquired a representative, legislative assembly *(Duma)*, elected by indirect suffrage. However, the executive remained accountable only to the Emperor who could invalidate any of the laws approved by the Duma. In 1906 Nicolas II signed a decree which allowed for free access to the civil service of all citizens regardless of their social origin.

Russia joined World War I in 1914. The strains of the war led to a breakdown of the country's political system and social fabric. Food riots in St. Petersburg and other cities toppled the monarchy in March 1917. The provisional government established by the Duma was short lived and dissolved when the Bolsheviks, headed by V..Lenin, seized power in November 1917.

The Soviet Government

The revolution in 1917 brought about dismantling of the old state and governance structures. A decree of the Soviet government of 11 November 1917 "On the Abolition of All Estates and Service Ranks" (Grishkovets, 2002:71) cancelled all civil service ranks as well as formal division of employees into those hired by the state and those working in the private or cooperative sectors. Many of the old-regime cadres either left the country or refused to co-operate. They were replaced by new servants selected mainly in accordance with criteria of origin and ideological fidelity. As a result most vacancies in newly created Ministries (*Narkomaty*) and Councils (*Soviets*) were filled with people whose professional competence and performance were questionable.

The original Marxist concept of building an absolutely new state apparatus in line with the slogan that every kitchen maid could manage the state shortly proved its absolute fruitlessness. The bolsheviks soon realized that they had to make use of "old" specialists and train "new". *As a result, instead of introducing radical changes into traditional administrative and behavioural culture the communist regime built upon it.* The state apparatus increased quickly in size and adopted the worst traditions of tsarist bureaucracy. V. Lenin realized this, albeit too late, when he could no longer influence the events being encircled by a hostile environment created by I. Stalin and immobilized by sickness. He wrote in one of his last letters: "If we take Moscow with its 4,700 communists in responsible positions, and we take the huge bureaucratic machine, that gigantic heap, we must ask: who is directing whom? I doubt very much whether it can be truthfully said that the communists are directing that heap. To tell the truth, they are not directing, they are being directed". (Lenin, vol.33, page 288).

The evolution of the Soviet system of governance led to the creation of a rigidly bureaucratic and inward-looking administrative apparatus, performing strictly controlled functions and having a limited capacity for decision-making. The whole structure rested upon the Communist Party's Politburo and Central Committee for its policy formulation, decision-making and overall performance review. The structure functioned from the centre with a network of Party Committees in republics, oblasts, towns, districts right down to the level of a factory, school, office etc.

A typical feature of the Soviet governance structure was strict vertical subordination of all levels of authority with very weak, if any, horizontal links and no possibilities of independent decision making at levels below the Party's Central Committee or a Ministerial Board. Decisions were usually taken at the very top and all elements of a 'pyramid' were obliged to exercise them. The party 'functionaries' were present in every large office and controlled the activities and conduct of employees and compliance of management decisions with Party directives. Membership of the Communist Party was a minimum requirement for promotion to any post of significance in any state or public organisation. In general, the state apparatus in the Soviet Union was inefficient and accountable only to the Party top leadership. It consumed tremendous public resources and lacked the confidence of ordinary citizens.

Though the rule of law, representative institutions, direct participation, individual rights formally existed, in reality the political leadership, which in fact was made up of top party and state bureaucrats, exercised absolute power through party representatives and bureaucratic centralism. This power was complemented by a system of privileges and formed the base of a kleptocracy called the "nomenklatura".

In the Soviet Union there was no formally constituted civil service as a legal and social institution. Employees of the government and different levels of state administration did not have a special status distinguishing them from the rest of hired personnel. Since over 95% of the industrial sector and that of provision of services belonged to the state almost all the employees were considered to be at the service of the state. Accordingly, the government had no central agency charged with the management or supervision of the civil service. Performance based career advance in its modern understanding was virtually absent. The Party organs decided which individual is suited best to different type of work on the basis of loyalty, political consciousness, knowledge of the tasks and organizational abilities. The state administrative elite made up an essential part of the communist party top bureaucracy and played an important role in the state policy formulation and implementation throughout the whole Soviet period. A Western researcher described the Soviet Union at the time as an "absolutist bureaucratic state" (Robinson 2002: 43)

Perestroyka

The changes initiated by M. Gorbachev in the late 1980s and referred to later as "perestroyka" brought about fundamental transformation of the political and economic system of the Soviet Union, led to its break up in 2001 and formation of *new independent states* on the territory of the former Soviet Union. The Russian Federation inherited the bulk of the state apparatus of the former Soviet Union. Already in 1989 the Supreme Soviet, the country's highest legislative organ, was considering draft legislation that provided for the establishment of the Russian civil service. These plans were not realized at the time due to political turbulence and overall instability in the country.

In summary, we may conclude that typical of Russia throughout most of its history was a special mechanism of formation of the political elite through the service to the state. As a rule the "list" of Russian political elite comprised a considerable part of top-level administrative bureaucracy. At the same time top and mid level state bureaucracy depended almost entirely on a will of a monarch or "*politburo*". Such "construction" encouraged the formation of patron-client relations and development of highly politicized and centralized state administration.

It is noteworthy that in his work "The Parliament and Government in the Transformed Germany" Max Weber assessed the administrative systems in Germany and Russia in early 20th century as "the rule of bureaucrats" (Beamtenherrschaft) (Maslovsky, 2001:25). This rule is characterized by a considerable degree of appropriation of the policy-making function by bureaucrats. In M. Weber's view this tendency is present in any bureaucratic organization but it develops notably only when effective control over bureaucracy is lacking.

Although the history of Russia is predominantly a history of non-democratic forms of governance, it also incorporates many instances of broad but slow reforms (more or less successful) affecting to a greater or lesser extent the state administrative structures. As a rule reforms were initiated from the top under the influence of either internal (contesting opposition pressures, unrest and turmoil, territorial expansion...) or external (wars, cultural influence…) factors.

3.3 Constitutional Framework

In accordance with the Constitution adopted on December 12th 1993 the Russian Federation is a federal, presidential, democratic republic. The executive power is exercised by the Cabinet of Ministers, legislative functions by the bicameral Federal Assembly and judicial authority by the RF Constitutional and Supreme courts. The Article 10 of the Constitution of the Russian Federation states, that "organs of legislative, executive and judicial power are self-dependent". The President is the Head of State and serves mainly as the guarantor of constitutional order. He also defines major directions of internal and foreign policies of the state and ensures coordination and interaction of activities of all branches of power. In practice the President performs substantial executive functions.

The Constitution provides for wide powers in the President, who appoints and dismisses the Chairman of the Government (Prime Minister), Members of the Cabinet, top ranking State officials and Armed Forces officers, has the right to dissolve the State Duma and order new elections. Although the Parliament approves nomination of the Prime-Minister and some high-ranking officials, in reality the President can use many devices to enforce his will upon the Federal Assembly.

According to the new Constitution (in contrast to the previous one adopted in 1991) the powers of the Parliament to exercise effective control over the Government have been considerably reduced. A number of Ministries and Agencies have dual subordination – to the President and to the Cabinet of Ministers. The constitutional division of powers between the president, the executive and the legislature has not been properly defined yet.

The state structure of the Russian Federation comprises three basic levels: 1) the Federation; 2) the Subjects of the Federation (regions); 3) the territorial entities which make up the Subjects of the Federation and which form the basis of local self-government system.

Article 65, para.1 of the Constitution stipulates that the Russian Federation is composed of 89 Subjects of varying legal status (reduced recently to 87 Subjects). They differ from each other in population, territory and economic potential. However, firm criteria for defining the status of Subjects and distribution of authority between a Subject and the Federal Centre have not been defined. The Subjects of the Federation have their own legislative, judicial and executive bodies (administrations), the latter headed by a President or a Governor.

The local self-governing authorities operate at the level of settlements, municipalities or districts in cities or large towns. Until recently the extent of self-governing authority was determined by each region (Subject) since the Law provided for some flexibility in this respect (whether that is good or bad is a separate issue). In general self-governing authorities and central government act relatively independently from each other. However, some Subjects of the Federation established special branches of state administration at the level of municipalities and settlements which means that there exists a struggle for competence and authority between separate regional administrations and self-government bodies.

A large number of Federal Ministries and Committees have their branches or representations in the Subjects of Federation (regional level) dealing with such matters as enforcement of federal law, management of federal property, maintenance of unified customs and taxation regulations, support for federal energy and transport systems, etc.

As a rule both regional and federal authorities have joint jurisdiction over ownership of land, water and natural resources, educational and health-care institutions, and social protection systems. There are also areas where the regional authorities are quite self-dependent. Civil service related Laws from 1995 and 2004 stipulate that organisation and management of the civil service of a Subject of Federation rests under its authority, whereas development and approval of the legal base for civil service activities remains a matter of joint jurisdiction with the Federation.

3.4 Political System

The peculiarities of the political system and political regime are of over riding interest as factors which have direct impact on the development and change of civil service systems. Following the demise of the communist system and dissolution of the Soviet Union the new political leadership voiced its commitment to the key principles of democracy and many of these principles became part of usual practice.

Throughout the 1990s the Russian political system was evolving with occasional disruption by periods of instability or even crisis. The fragility of Russian democratic institutions was evident, for example, in the long- standing discord over the balance of power among the three key state institutions: the President's Administration, the Government and the Parliament. This discord culminated in October 1993 when military troops loyal to the President stormed the building of the Russian Parliament.

A distinguishing feature of the Russian constitutional and political system is that the government is formed not by a winning party/coalition, but by a Prime-Minister appointed by the President. The Constitution does not oblige the President to select candidatures for the position of the Head of Government from only those representatives of the party/coalition which have won elections.

The process of policy-making in Russia is rather complicated and is subject to excessive personalistic influences. In this connection it is difficult to disagree with those observers and experts who pointed to obvious authoritarian tendencies of the Russian political leadership and defined the existing political regime, in particular during the times of President Yeltsin, as corporatist-autocratic.

Notwithstanding recent changes in the legal framework, the authority, responsibility and reporting obligations of the executive have not been sufficiently defined and formulated yet. The RF Constitution and other legal acts provide for dual authority - of the President and of the Head of Government - over the executive power, and this creates duplication and lack of co-ordination. For instance, the President exercises direct supervision over "law enforcement" agencies and some others – Ministry of Defense, Ministry of Interior, Ministry of Foreign Affairs, Security Service - although their heads are members of the Government.

The system of political parties in Russia remains largely underdeveloped. It can hardly be called representative and serves mainly to accommodate the interests of various lobbies. As one of the President's assistants put it "In essence we do not have political parties" (Otechestvennyie Zapiski, 2004:95). The rule of law has been declared, but its occasional abuse in particular by top politicians and government officials largely undermines its practical value. Furthermore, leaders in post-communist Russia do not deem themselves bound by constitutional provisions they find inconvenient or dislike. Arbitrary postponement of the Presidential elections to June 1996, as well as large-scale financial machinations during this elections campaign serve only as one out of many examples.

The political and economic environment in Russia characterized during the recent decade by weak law enforcement, developed patron client relations, embedded corruption and immature civil society appears to have been ripe for abuse by politically connected groups. This period also witnessed gradual devolution of economic and political power from the federal to regional and local authorities. To address region-specific concerns and diffuse potentially disintegrating trends, Boris Yeltsin concluded bilateral power-sharing agreements with many of the 89 Subjects of the Federation which comprise the Russian Federation. So Russia in the 1990s became a much less centralised state than it was during Soviet times.

Boris Yeltsin's resignation on December 31, 1999 and the election of President Vladimir Putin after the first round of voting on March 26, 2000 marked the first democratic transition of a leader in Russia's history. Vladimir Putin appears to have come to office with the main objective of consolidating central authority. He also voiced his commitment to reduce the powers of the Russian oligarchs and to reform the Russian bureaucracy.

A series of measures adopted after the election of President V. Putin appear to have strengthened the central executive power, seriously undermined the powers of the oligarchs and consolidated dominant positions of the centrist parties which largely represent the interests of state bureaucracy.

3.5 Economic Situation

The economic development of the Soviet Union was characterised by centralized state planning, over-investment in heavy industry, low labour productivity in virtually all sectors of the national economy. There was underdevelopment of the service sector and light industry, unbalanced regional development and wasteful energy-consumption practices.

The economic reforms launched in Russia in the early 1990s followed a radical-liberal conception of market transformation said to be prescribed by IMF in line with the policies known as "Washington consensus" and often referred to as "shock therapy" concept.

Although the reforms laid down the foundations of the market economy they also resulted in a sharp decline in national income, impoverished most of the population, led to sharp decrease in life expectancy, and contributed to growth of corruption and organized crime.

Senior Vice President and Chief Economist of the World Bank Joseph Stiglitz in the lecture given at the annual session of the World Institute of Development and Economic Research (WIDER) in Helsinki in January 1998 said, that "the set of policies which underlay the Washington Consensus are neither necessary nor sufficient, either for macro-stability or longer-term development." He declared that these policies are "sometimes misguided", that they "neglect ... fundamental issues", are "sometimes even misleading", and "do not even address ... vital questions." (Bogomolov, 2000:4)

In this connection one of the leading Russian economists wrote "In the years of radical reform, Russia lost its erstwhile economic might, falling in terms of Gross Domestic Product (GDP) even behind such countries as Mexico, Brazil and Indonesia, with a five-fold lag behind China and a ten-fold lag behind the United States. The recession continued, year after year, despite government declarations to the contrary. As compared with the pre-reform level, the productivity of the Russian economy on the threshold of the new century was more than cut in half, industrial output fell to 40 per cent of what it had been, and in the light industry and the food industry it dropped by two-thirds". (Bogomolov, 2000:2)

The devastating crisis in 1998 resulted in a situation when Russia defaulted on its debts and devalued its rouble by almost 75%. Poverty expanded further and remained one of the main obstacles for the country's progress. According to OECD data, between a quarter and a third of Russians live below the official poverty line, and this situation causes "continuing disturbing trends in some demographic and health statistics" (OECD, 2006:1).

The main features of the economic situation throughout the 1990s were huge external debt ($ 145 billion on average), continuing slump in production and investment, and growing unemployment (12,4% in January 1999 - official statistics). Serious financial constraints forced the Government to cut even the most essential social programmes and to hold down the growth of salaries in the public sector. As a result remuneration of civil servants was downgraded to the level where in many cases it ceased to serve as a motivating factor.

Overall, the situation seemed to have notably changed for the better starting from 2000. The OECD noted Russia's economic growth of 15% in 2000-2002. "The Russian economy has experienced a number of favourable trends and developments," the OECD report said. "Output, employment, consumption and investment have grown significantly." The OECD praised the Russian government's efforts in eliminating the chronic budget deficit and political stability that has enabled the Kremlin to push some crucial reforms through parliament. Many observers attributed this improvement to three major factors - high oil and gas prices, cheap national currency and important reforms implemented in the past two years.

Other economists were much more reserved in their assessment. "Although the years 1999-2001 appeared to be the best for the Russian economy witnessing GNP growth by 19.7%, industrial production by 30.4%, and investment by 33.5% the subsequent period proved that the growth was not sustainable. The industrial growth slipped from 11.9% in 2000 to 3.5% in 2002. The rate of investment growth has fallen four times for the same period". (Delyagin, 2002:11)

The unofficial drain of capital, which leaves the economy virtually bloodless, remained almost at the same level of USD 22 billion in 2000 and USD 21 billion in 2001 (Delyagin, 2002:13). The number of unemployed amounted to 8.8% of the labour-capable population in 2001. The capital drain was caused not so much by inflation or by taxes, as by weak if any protection of property and the abuses of the state bureaucracy whose "bribe taking capacity" has increased two fold (assessment by businesses) during the period 2000-2002" (ibid).

Sharp increases in the price of oil on the world market in the early 2000s created favourable conditions for the flow of considerably higher revenues to the state budget. Proficit of the federal budget equalling 1.5% of the GDP in 2000 had grown to 7.4% of the GDP in 2005. (Alliance Media, 3 April 2006).

3.6 Privatisation

The establishment of new political and economic institutions in Russia in the 1990s was accompanied by massive redistribution of state assets: this facilitated administrative corruption. New horizons for embezzlement and graft opened up with the launch of privatisation. While appropriate legislation, control and accountability were virtually absent, resource allocation had remained in the hands of government officials. They, in most cases, unilaterally decided which strategic enterprises were to be privatised and what were the rules of the divestment.

What happened in reality was transformation of political (party/bureaucratic) power into economic/financial power and material wealth (realty, securities etc). Mass redistribution of property during the initial voucher privatisation and the loans-for-shares scheme of 1995-96 was marked by so many violations of the far from perfect laws and rules that any decent decision in this respect was regarded as an obvious exception.

Distribution of property rights to a privileged group of people (as a rule a narrow group of former directors, party "functionaries" and loyal relatives/acquaintances) contributed to the concentration of economic power in the hands of influential oligarchs. From 1995 onwards privatisation was frequently achieved through the sale of state assets at below market prices to financial-industrial groups with connections to the government. Later these groups pressed the state for special privileges and opposed reforms aimed at creating a more competitive environment. (Rakityansky, 2002:20).

Privatisation has not brought the expected revitalisation of the private sector since it was not accompanied by an appropriate regulatory reform, changes in management and in the surrounding environment. As a rule the new owners preferred to squeeze immediate profits from their new property instead of investing in its long-term potential.

The lack of transparency and numerous violations in the privatisation process have undermined further public confidence in the government's ability to manage the economy in an impartial manner and raised concerns over widespread corruption. The interests of influential business corporations were often (especially during the presidency of B.Yeltsin) equated to state interests and priorities and a large part of the cadre of senior civil servants acted as lobbyists of oligarch interests.

Even after the privatisation of the 1990s, about fifty percent of the public wealth remained under state control. "It is managed "amazingly poorly," said former Deputy Minister of Economy M. Dmitriev. "But even an efficient public administration could not successfully manage so much property. The state presence in the economy must be reduced, and the public administration must be streamlined to manage better what remains under state control". (Dmitriev, 2004)

3.7 Legal Framework

In the times of the monarchy one of the principal normative acts that regulated the whole range of administrative service relations was the Civil Service Charter adopted in 1886. During Soviet rule a separate legal and normative base covering activities of civil servants and their status did not exist. Instead, provisions of the Labour Code and in-service regulations of each state agency were applied.

The Constitution of the Russian Federation refers to the term "civil service" only three times – in Article 32, point 4, which says that "citizens of the Russian Federation have equal access to civil service employment"; in Article 71, point t) which says that "the federal civil service comes under authority of the Russian Federation" meaning that it shall be a unified service across the territory of Russia; and in Article 97, point 3, which says that "deputies of the State Duma shall not be acting civil servants".

The first legal documents that initiated the formal origin of the Russian Civil Service in modern times were the Presidential decrees "On Priority Measures to Organize a System of Civil Service in the Russian Federation" and "Regulations on the Federal Civil Service" signed by the President in 1993 respectively on 3 June and 22 December (Atamanchuk, 2002:63).

The first basic legal act, which dealt with civil service issues in a comprehensive manner was the Federal Law "On the Basic Principles of Civil Service in the Russian Federation" (enacted in July 1995).

The Law stressed "the priority of the Constitution of the Russian Federation and the federal laws over other regulatory legal acts and service instructions related to the exercise of official functions by the civil servants and enforcement of the rights thereof" (Article 5, point 1) and provided for "mandatory compliance of all civil servants with the decisions taken by higher state bodies and authorities within the jurisdiction thereof and in keeping with the legislation of the Russian Federation" (Article 5, point 6).

The law defined the legal status of civil servants; their employment rights and duties; determined recruitment and promotion procedures; envisaged arrangements for civil service management; introduced a disciplinary code and a code of ethics, declared "non-partisanship" of the civil service as one of its basic principles (Article 5, point 11).

The law forbade civil servants in particular to be engaged in:

- all paid work (except pedagogical and creative work);
- 'commercial activity' or lobbying;

- misuse of state property;
- the receipt of gifts in connection with the exercise of their duties;
- strike action;
- leaking of confidential information;
- employment of relatives as subordinates.

Overall the law provided for the introduction and enforcement in Russia of the principles of the classic 'Weberian' model of civil service. However, a package of subsequent normative acts (around 30) envisaged in the Law never materialized. Instead, a dozen Presidential decrees and Government enactments on various aspects of civil service activities were signed. However these legal documents did not proceed from a common concept.

The fragmentary and often uncoordinated nature of legislation concerning the civil service at that time manifested itself for instance in the fact that such fundamental legal acts as the Law "On the Government of the Russian Federation" (approved by the State Duma on April 11th 1997) and the President's Decree "On the System of Federal Bodies of Executive Power" of August 14th 1996 did not include any reference to a civil service!

The situation started to improve in the early 2000s following election of the new President. The legislation regulating the civil service proper was considerably enhanced with the adoption of two new laws "On the System of Civil Service of the Russian Federation" (adopted on 27 May 2003) and "On the State Civil Service of the Russian Federation (adopted on 27 July 2004). A brief analysis of these laws will be carried out in the following chapter devoted to the reform of the Russian civil service.

Further, in the period 2000-2005 the government ensured passage of legislation which addressed in one way or another the issues of corruption and efficiency of state administration and artificial barriers for entrepreneurial activities.

The legislation included laws and other normative acts on:

- deregulation targeted at facilitating registration and licensing procedures and reducing the number of inspections and certificates required of enterprises;
- tax and customs administration reducing significantly opportunities for arbitrary actions by the authorities;
- judicial reform in particular delineating roles of the various actors in the legal process and creating barriers for graft;
- labour relations (labour code) eliminating some rigid constraints on employer-employee relations;
- greater transparency and accountability of the administrative apparatus.

Although the legal and regulatory base for the civil service has been recently considerably improved as part of civil service reform efforts the institutional framework for implementation and enforcement of legislation remains weak. Overall the judicial system has not yet developed sufficiently and to the extent necessary to embed the rule of law. That requires among other things a change of mentality and practice. Many legislative acts have never been fully implemented due either to a lack of financial resources, low respect for the law or due to resistance of the bureaucrats affected by the novelties.

At the same time laws at the regional level sometimes contradict federal legislation. The sections of administrative law regulating civil service are not properly developed and the establishment of the system of as yet non existent administrative courts is in the very initial stage.

An important feature of the normative environment in Russia is the internal in-service regulation adopted by each ministry or agency. Although these internal instructions and rules do not always fully comply with the federal legislation, they are regarded by most civil servants as having greater 'weight' for their work compared to federal laws. Once again it should be stressed that in Russia the adoption of regulatory norms and rules and their practical implementation and strict observance are often separated by considerable time delay.

3.8 Central Government

Throughout the 1990s the system of governance in Russia largely retained the overly centralised and strictly hierarchical 'command' structures typical of the Soviet period. Although the number of federal ministries and agencies has been reduced from around 140 sector-oriented institutions (in 1989) to approximately 65 (in 1995) the composition proper of the Central Government has remained basically unchanged. As of September 2002 the Central Government comprised the following agencies: Federal Ministries (23), State Committees (13), Federal Commissions (2), Federal Services (13), RF Agencies (3), and Federal Inspectorates (2).

A large number of Federal Ministries have their branches or representations in the Subjects of Federation dealing with such matters as enforcement of federal law, management of federal property, maintenance of unified customs and taxation regulations, support for federal energy and transport systems, etc.

The administrative reform launched in 2003 led to further reduction of the number of ministries – from 23 to 16. Reorganisation of the Russian government in 2004 is analyzed in more detail in the following chapter.

The Government is the highest authority of executive power. The Law on the Government of the Russian Federation adopted in April 1997 (amended in December the same year) defines the basic functions of the government. In accordance with the law the Government:

- determines and implements home and foreign policy of the Russian Federation;
- exercises regulation in social and economic spheres;
- ensures the unity of the system of executive authority in the Russian Federation,
- directs and supervises the activities of executive bodies;
- elaborates targeted federal programmes and ensures their implementation.
- draws up and implements the federal budget;
- manages federal property;
- carries responsibility for home security and defense.

The Government has the right of legislative initiative. Members of the Government (Cabinet of Ministers) meet formally at least once a week. Another distinguishing feature of the government system in Russia is the existence of influential Apparatus of the Government which has so far executed a sizable part of policy making activity of the line ministries.

The second half of the 1990s saw numerous reorganisations of the Executive branch at the federal level caused almost exclusively by political and severe budget deficit related factors. As a rule such operations were poorly organised and their implementation left much to be desired. Usually the planned downsizing of various ministries, departments and committees shortly resulted in the growth of the number of employed staff.

Typical of the overall system of governance in Russia has been focus on inputs rather than outputs, functional overlaps, disputed division of responsibilities and rivalry between ministries and agencies.

Decision-making at the level of central government has been flawed by poor demarcation of tasks and responsibilities between the President's Administration and the Federal Government, between federal agencies, between central and regional governments and has been lacking capacity for sound policy analysis. Control over implementation of Government decisions is not effective, and accountability for poor performance is weak. The system of regular forecasting and surveys of public opinion on specific issues is undeveloped. Strategic plans for development of separate ministries and agencies have been the exception rather than the rule.

Each ministry elaborates on its own state policy in its tasked area and most importantly usually without prior or post consultations with parliament, political parties, professional organizations, or representatives of civil society.

The existing administrative culture gives preference to a system of decision-making that is far top-heavy. Many unimportant decisions are taken to the highest levels in the state apparatus resulting in slow, inefficient response and red tape. As a rule little or no explanation is given to the public why a particular decision was made and what benefit it would bring.

The administrative reform launched in 2003 was designed to deal with all these problem areas. The extent to which it succeeded or failed is analyzed in the following chapter dealing with the reform of the Russian civil service.

3.9 Civil Service: Status, Efficiency and Institutional Framework

After the dissolution of the Soviet Union, the Russian Federation inherited an over-centralised, ideologically driven system of state administration and a corps of state administrators governed by a combination of modern and outdated rules, traditions and informal relationships, some of them dating back to the times of monarchy. As an institution the Russian civil service has been in the process of formation since the early 1990s. The impetus to its establishment was given by the President of the Russian Federation in his decrees of 3 June and 22 December 1993 announcing the creation of the Federal Civil Service and mapping out the principles of its organization. This process was further consolidated in 1995 when the Parliament adopted the Law 'On the Basic Principles of Civil Service of the Russian Federation'.

In spite of some obvious deficiencies, for the first time in the modern history of Russia this legal document gave an official definition of the civil service.

The document laid down the legal basis for its organisation, outlined the procedures of service and introduced Job Classification System as well as distinction between political appointees and career civil servants.

Throughout the 1990s the formation of the civil service was taking place within the institutional framework which was heavily affected by the heritage of the Soviet times. In line with the principles of federalism and existing territorial structure the Russian civil service is divided into federal and regional services (Subjects of the Federation). Formally, local self-government and hence municipal service are not regarded as part of the system of state management although in practice they are largely financed from the budgets of the state and the Subjects of the Federation.

The Russian civil service at both the federal and regional levels is further divided into civil services of the legislative, executive and judicial branches of power. The new law "On the System of State Service in the Russian Federation" adopted in 2003 introduced additional differentiation of the entire state service into civil (public), military and law enforcement services.

At present the Russian civil service does not constitute a unified and homogenous system yet. In practice the civil services in the Subjects of the Federation (regional level) are governed by norms and rules that do not always comply with the Federal Law. A single management body or unit for the whole civil service, or at least for its federal level, is not yet established. The service lacks stability from both the institutional and human resources points of view. The legitimacy of the Russian civil service is undermined by lack of clear definition of its place in the context of institutional roles and division of powers existing since the 1990s.

The efficiency and effectiveness of the working of state administration is widely acknowledged to have dropped considerably compared even to the Soviet period. Partly this can be explained by overall weakness of the state and frequent inability of state institutions to perform many of their basic functions even up to the lowest standards. A further explanation is frequent underestimation of the crucial importance of a sound institutional and regulatory framework and efficient state administration for the country's prospective development.

It is estimated that Russia loses up to 2 % of its annual GDP because of the poor system of public management. Over third of interviewed civil servants in 2000 consider that effectiveness (efficiency) of their work is low (RACS, 2002:11). The Chief Control Department of the Presidential Administration reported in 2002 that almost half of the Presidential orders and instructions given to the Government were not implemented on time (Nikolaev, ed. 2006:116).

The post communist period saw the appearance of further negative trends that affected the development of the civil service in Russia:

- civil servants' pay declined sharply compared with salaries in the private sector and even in comparison with average incomes in the country;
- although the communist party rule was abolished the civil service remained highly politicized causing among other things a considerable degree of instability of the personnel;
- overall professional level of the civil service corps declined, in particular, due to exodus of most experienced and qualified staff to the private sector;

- enforcement of civil service legislation and other normative acts remained weak.

The turbulent political environment, frequent reorganisation of ministries, high turnover of cadres, low (until recently) prestige of the service, insignificant remuneration, and overall atmosphere of uncertainty hampered the identification of values that could define and cement a common sense of mission held by most members of the civil service.

Besides, measures adopted to create foundations for the development of a modern, efficient, accountable and corruption immune civil service did not proceed from a single concept agreed among major stakeholders. By the mid 1990s a model of career based civil service gained a foothold in Russia with many elements borrowed from the French civil service system.

3.10 Civil Service Management

As can be seen from preceding subchapters the civil service system is a complex organization made up of various elements tied together by multilateral links and relationships. Therefore, management of this organization and coordination of its elements is essential for effective functioning of the whole system.

From the early days of the establishment of modern Russian civil service there has been no effective central agency/body in charge of its management. Although article 26 of the law "On the Basic Principles of the Russian Civil Service" (adopted in 1995) stipulated, that "coordination of activities connected with the handling of the tasks ensuing from this Federal Law shall be effected by the Council on the Civil Service under the President of the Russian Federation", this council never performed the management function. Furthermore, the Council was convened very irregularly, and soon ceased to exist.

In practice, resolution of many issues related to the formation, management and normal functioning of the federal civil service were either distributed among key ministries and agencies (e.g. cross cutting issues of defining civil servants basic payment and funding of their in-service training were dealt with by the Ministry of Labour), or were the responsibility of the leadership of the respective parent ministry/institution. Sometimes the administration questions were simply unaddressed for many years. In the Subjects of the Federation almost all civil service related issues were resolved by the regional authorities.

Generic issues related to the overall development of the civil service were usually addressed by the Duma, Presidential Administration and the Office of the Government.

Management of various civil service units was affected by the peculiarities of the political and economic situation as well as by existing administrative traditions and culture. For example, numerous financial, human resource and circumstantial constraints often distracted attention from the issues of strategic management which were either ignored or attracted a secondary importance. The overall impact of management was reduced by excessive concentration on control and very limited attention if any to dialogue and participation.

Additional complexity stemmed from poorly defined and often overlapping authority, duplication of functions and competencies between various executive agencies. Excessive centralization of decision-making necessitated lifting of too many issues to a very high level in the system for resolution.

Article 16 Chapter 3 of the new Law 'On the System of State Service' (adopted in 2003) provided for the creation of a Federal body in charge of managing the federal civil service and respective bodies in the Subjects of the Federation in charge of regional civil services. Thus, the law did not envisage a central management structure for the entire civil service. Such a structure has been regarded by many observers as one of the preconditions for the integrity and effectiveness of the Russian state administration. Furthermore, five years after approval of the law such a body at the federal level had not been established yet.

3.11 Relations between Civil Servants and Politicians

Complete subordination of the administrative agents to the political authority is deeply rooted in Russia and dates back to the period of the establishment of unified Russian state. Although traditionally the state bureaucracy in Russia used to regard itself as a body serving the interests of the ruling elite, the latter in fact was in part composed of top representatives of that bureaucracy.

The nature of the relationship between the politicians and the state bureaucracy in Russia has hardly changed as compared to the communist past and even the pre-revolutionary (1917) period. These relationships have lost their ideological imperative but as in the past the bureaucracy remains an obedient tool in the hands of the political elite. Simultaneously the bureaucracy tries to impose its corporate interest on society and to present it as the latter's common interest.

The essence of such a relationship was defined by Niskanen who, although not referring specifically to Russia, characterized the power relations between bureaucrats and top politicians as a"bilateral monopoly" (Niskanen, 1971:56). However in his opinion, in practice the bureaucrats could generally pursue their interests and achieve their own objectives.

Although the civil service law from 1995 established formal distinction between political and career posts this differentiation was not always clear-cut in practice. Throughout the 1990s and later, Russian senior civil servants as a rule had partisan links with the ruling political elite. It was common practice for covert and in some cases even undisguised involvement of civil servants in the election campaigns of governors, mayors, and heads of administration. Where there were changes of party or coalition in power or government, changes took place in the civil service cadre. "The political nature of administration and dependency on individual directors make the position of civil servants unstable..., career officials are also often replaced on political grounds". (Larjavaara, 2001:15). Replacements of top and mid level civil servants following elections have been particularly widespread at the regional level (Otechestvennyie Zapiski, 2004:54).

It is noteworthy that the requirement for civil servants' non partisanship imposed by the law adopted in 1995 was later dropped in the civil service law approved in 2003. Although Article 4 of another civil service related law (2004) defined one of the civil service guiding principles as "protection of civil servants against undue interference in their professional service activity", this principle has never really worked yet. Similarly, very weak enforcement is typical of another clause stipulated in Article 17 of the same law. With respect to civil servants this clause prohibits "use [of] official powers to the benefit of political parties, other public or religious associations and other organizations, or publicly express[ing] personal opinions in respect of such associations and organizations in his/her capacity as a civil servant, unless envisaged by his/her job responsibilities".

Political administrative relations in Russia have also been affected by absence of serious political competition among main political parties and coalitions. Influential political rivals were violently suppressed by President Yeltsin in 1993 and since then there has been no strong political opposition in the country.

The position of the bureaucracy versus politicians appears to have strengthened in the 1990s. The following example demonstrates the ability of the administrative apparatus to effectively ignore decisions of the political leadership in particular when they run counter to the interests of the bureaucracy. The President's Decree of 1992 "On Fighting Corruption in the Civil Service System" stipulated that all civil servants should regularly submit declarations of their income and assets. However, it took five years and required another President's Decree in 1997 to enforce practical implementation of this requirement even on a limited scale.

The bureaucracy continues to be a major player in selecting and supporting candidates for key political posts. In Russia the strength and influence of a politician are determined to a considerable extent by the number of high-ranking supporters he has in key administrative positions. A recent OECD report called Russia "a weak state with strong bureaucrats" (OECD, 2006:1)

The formation of transparent and permanent rules guiding relations between politicians and civil servants is complicated by virtual absence of clearly defined constitutional roles for various institutions and agencies, by lack of long standing democratic traditions and above all by the patronage culture prevailing within the administrative establishment's networks.

3.12 Human Resource Management

According to data provided by the Federal Service of State Statistics the total number of civil servants and other employees of federal state bodies, state bodies of the constituent entities of the Russian Federation, and other state bodies established under the Constitution of the Russian Federation (hereinafter collectively referred to as the state bodies), together with municipal servants in 2002 (i.e. close to the start of civil service reform) totalled 1,252.3 thousand, including a verified 315.1 thousand civil servants and other employees of federal executive bodies. During the period of 1992 to 2002, the number of civil servants and other employees of these bodies grew by a factor of 1.8 (Federal Programme "Reform of the Russian Civil Service", 2002).

By 2007 the human resource base of the Russian civil service has increased further and reached 1, 624 thousand persons. The evolution of the number of employed personnel in the Russian state bodies since 1992 (following break up of the Soviet Union) is shown in Table 3.1

Table 3.1 The Cadres of the Organs of State Authority in the Russian Federation*

	1992	1993	1994	1995	1996	1997	1998	1999
Overall number of civil servants (thousands)	816.1	921.6	1004.3	1061.8	1093	1108.9	1102.8	1133.7
	2000	2001	2002	2003	2004	2005	2006	2007
Overall number of civil servants (thousands)	1163.3	1140.6	1252.3	1300.5	1318.6	1462	1577.4	1623.9

Source: Annual Statistical Bulletin 2005 & 2007, Federal State Statistics Service; "Russia in Figures 2008", Federal State Statistics Service.

* The data from the Federal State Statistics Service and other sources (e.g. Federal Programme "Reform of the Russian Civil Service", referred to here and in the following chapters) may differ. This can be explained by the fact that apart from civil servants proper these data may comprise also support and technical staff employed by the state bodies.

Compared to state administration of other industrialized countries the Russian civil service does not seem to be overstaffed. In 1999 Russia had 2.8 federal civil servant per 1000 employed population, whereas the same ratio for USA was 3.9, and for UK 5.2. (Federal Programme "Reform of the Russian Civil Service", 2002).

Typical of the Russian civil service, throughout the 1990s but especially in its first half, was a high level of personnel turnover. For example, 58% of cadre in the federal executive bodies and 30% in the respective regional bodies were replaced in course of 1992-1993 (RACS, 2002:145). As a rule those who left were the best. By 1997 there had emerged an approximate numerical parity between civil servants with a record of government or party employment during Soviet period and those who joined the state administrative apparatus after the start of democratic transformation (Grishkovets, 2002:67).

From the very start of its re-establishment in 1993 the Russian civil service adopted a rather complex hierarchical structure of grades and ranks reminiscent of the Table of Ranks introduced by Peter the Great in the 18th century.

In accordance with the Law 'On the Basic Principles of the Russian Civil Service' (approved in 1995) all civil service positions were classified into categories (A,B,C) and groups (1-5). Further, all civil servants within each of five groups (supreme, key, major, senior and junior civil service positions) were assigned qualification ranks (five ranks all together) of the first, second and third class based on the results of a state qualification examination.

Holders of Category A, civil service position, were considered to be political appointments and formally not part of the civil service. Category B position was established for directly facilitating the exercise of the authority of category A position holders. Therefore civil service in category B posts was limited to the term for which the respective holders of category A posts were appointed or elected

(Article II, point 2). The most numerous group of civil servants are holders of Category C positions, who are tenure (career) civil servants.

The Law 'On the State Civil Service of the Russian Federation' adopted in 2004 introduced certain changes into the classification of civil service positions. Instead of categories A,B,C the law established four categories of civil service positions: executives, advisors, specialists and support specialists. The existing differentiation of civil service positions into groups and the existing system of ranks and classes were preserved. Political appointments were no longer associated with holders of civil service positions since they now formed a special group - holders of state positions (posts).

All civil service positions are included in the Roster of Civil Service Positions which is approved by the President of the Russian Federation. The roster of federal level civil service positions has existed since 2006.

In the Soviet Union employment conditions of government personnel were regulated by the Labour Code, complemented in some cases by agency specific (internal) regulations. The situation gradually changed during the 1990s together with adoption of civil service related legal and normative acts.

Since 1995 Russian civil servants are employed on the basis of contracts either with an indefinite term or a period of no more than five years. Each ministry continues to manage all its personnel except at the level of Deputy Minister and above. The case of the latter is the prerogative of either the Presidential Administration or of the Head of Government and its office.

Modern principles of merit based recruitment and career advance have been declared but their enforcement remains a problem. In spite of the fact that the civil service law from 1995 required holding competitive examination as part of recruitment and appointment procedures, in practice this has been rather an exception than a rule.

The mechanism of progression to the upper echelons of civil service is still predominantly very informal and comprises as its most important element the ability of a candidate to establish good personal contacts, to prove one's loyalty to superior management and to become associated with influential leaders or interest groups. Qualifications and competence have less significance "as it is personal loyalty that counts when it comes to appointments" (Nozdrachev, 1999:21).

At the same time the absolute majority of the middle and low ranking civil servants identify themselves with a corps of professionals and as a rule give no preference to any particular political authority or interest group in their everyday work.

Overall, modern principles of personnel management such as modern job descriptions, personnel information system, performance appraisal strategies and others are only nascent in the Russian civil service.

Domination of outdated principles of personnel management leads to various contradictions. For example, there appears to be a contradiction between those civil servants who are oriented towards making a career through highly professional performance and have relevant attributes and those who are motivated exclusively towards getting privileges and benefits through the mechanism of protection and patronage.

3.13 Payment and Incentive Policy

The prevailing in Russia system of motivation, compensation and promotion of civil servants is insufficiently linked to actual performance and achievement of tangible results. In practice and in most cases modern performance management is substituted by a system of benefits, bonuses and career advance which almost entirely depend on length of service and subjectivity of the immediate chief's opinion. Personal loyalty to the leadership and its disposition as well as "necessary connections and ties" are the most important factors which determine positive performance assessment of a civil servant.

In 2002, 59% of the interviewed civil servants expressed opinion that promotion based on personal ties and group interests (but not on merit) was the largest de-motivating factor impeding growth of professionalism and efficiency of the civil service cadre (Kotchegura, 2008:83).

In Soviet times, pay was only one of the elements in the system of civil servants' motivation since their status was largely derived from position in the hierarchy and respective access to various privileges (e.g. housing, foreign travel, special allowances, access to educational, health and leisure facilities, special shops). The present day situation may have changed in details and formalities but hardly so in essence.

Various types of bonuses, allowances and benefits continue to occupy a substantial part of the overall remuneration package for most civil servants. Remuneration consists of basic salary plus various extras for service rank, length of service, special conditions and bonuses. At present the total remuneration package amounts on average to 1.75 – 2.5 multiple on basic salary for low and middle level servants respectively and 4-8 times basic salary for top level bureaucrats.

Performance related pay has been declared one of the civil service reform's principal objectives but its mechanism is still in an experimental phase. Absence of direct linkage between achieved results and pay, and between merit and promotion prospects encourages preservation of underdeveloped service delivery standards and limited interest in user feedback.

As we already emphasized above, remuneration of an average Russian civil servant during the 1990s was unacceptably low. Average salary amounting to an equivalent of 200 USD in the mid 1990s increased to 400 USD in 1998 and then abruptly declined to 180-200 USD following the financial crisis and default of the rouble.

The situation with civil servants' pay began to improve in the early 2000s as a result of economic growth and sharp increase in state budget revenues following upsurge of international oil and gas prices.

According to data provided by Russian audit and consulting company FBK, only in 2002-2003 nominal salaries of civil servants were increased by a factor of 1.7 (Nikolaev, 2006:124). In 2004 the overall remuneration package including various bonuses and allowances was raised on average 5-8 times for Russian ministers and deputy ministers, 4 – 6 times for heads of departments and their deputies, 3-4 times for heads of section and their deputies, and 2-3 times for other civil servants (Vedomosti, 12 September 2004). It should be emphasized that the base-salary, transparent to the society and various control agencies, was increased at much slower rate in comparison with various types of bonuses.

The Federal State Statistics Service estimated an average monthly pay of a Moscow based federal civil servant in 2005 at the level of 24,124 rubles (approx. 800 USD). Average civil servants' pay in regional government across the country amounted to 16,612 rubles (approx. 570 USD). (Kommersant Daily, 27 March 2006).

3.13 Education and Training

Russia has a sound human resource base with a well-developed educational system that is capable of imparting a wide variety of skills, both basic and advanced. Before the fall of the communist regime state administration employees were re-trained and upgraded their qualification in the network of party schools and the Academy of National Economy.

In 1995 a Presidential decree established the Russian Academy of Civil Service with a wide network of regional branches. The central office of the Academy in Moscow offers training courses mainly for mid and top level civil servants.

In the same year the Russian Ministry of Education approved a standard of training for students enrolled on specialization N 0610 "State and municipal management". In 2002 this programme was delivered in 58 state and 12 non-state educational institutions of higher learning. The total number of students enrolled in this specialization was 20,000. (RACS, 2002:164).

However, the professional level of Russian civil servants in the 1990s and to a large extent at present is overall insufficient. For instance in 2004, in violation of the requirements of the federal law, 5.8% of civil servants in the category "advisors" and 25% in the category "specialists" did not have higher education. Among civil servants with higher education less than 5% graduated from "state and municipal management" faculties or departments. (Kotchegura, 2006:281)

Current legislation requires that every civil servant should undergo re-training or up-grading of qualifications at least once in three years. However, this requirement has never been fully met. For example, in 2004 only 15.7% of all civil servants passed through various types of training and up-grading of qualification (Federal State Statistics Service, 2005: 150).

The existing system of post-entry training of civil servants was frequently criticized throughout the 1990s and later. The expert community and mass media singled out the following main drawbacks:

- serious under-financing of training of civil servants;
- low relevance of training to the real needs and demands of civil servants;
- shortage of qualified teaching staff;
- weak competition between training institutions;
- no effective mechanism for identification of required skills and assessment of the training needs;
- very weak, if any, impact of successful training on promotion.

Among other weaknesses is shortage of specialized programmes for top level civil servants and of professional development programmes for every staff member. The enforcement of the latter could reduce fragmentation and inconsistency of training throughout a civil servants' career.

Starting from 2004 funding from the state budget of civil servants' post entry training has been increasing with overall positive impact on the quality and results of training programmes.

3.15 Societal and Administrative Traditions, Values and Culture

In exploring the essence of the Russian civil service system it is important to be aware of a long-standing tradition in Russia of domination of the state over society. The priority task in the 15-17th centuries was the formation of the centralised Russian state. Defending it against hostile neighbours required mobilisation of all resources leading to establishment of very close linkages between the state and the society. Consequently, all social groups had to provide direct services to the state and citizens were often regarded as "servicemen" of the state.

The role of the state and state administration was constantly growing under the monarchy and particularly during the Soviet times. Although the communist ideological postulates advocated disappearance of the state but in reality the Bolsheviks created a system where the state dominated all spheres of social life and controlled to a substantial degree the life of individuals.

Soviet state and society were characterised by a monopoly position of the communist party and excessive concentration of policy-making and planning functions in Moscow. Traditions and institutions associated with democracy and the rule of law were either weak or absent. The tsar's and communist bureaucracies were largely responsible for the widespread alienation of citizens towards the state administrations and for their cynical attitude towards the law.

Due to the dominant role of the state in the past and prevalence of entrenched patronage networks, the state is still considered by many citizens and civil servants as the main actor in resource allocation. The state is often perceived as a patron dealing with a select clientele. This is one of the reasons why many civil servants find it difficult to draw a clear distinction between public and private domains.

Many experts emphasize the uniqueness of the Russian culture in many ways different to European and Asian models (Kulinchenko, 2002). In contrast to the West European culture, Russian culture is less rational and pragmatic and is more emotional. It is less individualistic and more community oriented. Whereas western culture gives priority to efficiency and material wealth, Russian culture promotes domination of moral and spiritual values. At the same time this culture leaves enough room for the development of individual personality (e.g. world famous Russian scientists, writers, composers, artists). One of the main sources of Russian uniqueness is orthodox Christianity.

Orientation towards establishment of personal ties as the main instrument of solving various problems appears to be deeply rooted in the prevailing mentality. According to the findings of various representative polls up to 75% of the respondents consider good personal ties as an indispensable prerequisite for success in life (Otechestvennyie Zapiski, 2004:226).

Russia has a very limited (even when compared to the Eastern European countries) record of democratic bourgeois development but a lengthy historical legacy of absolutism and despotism. The monarchy was abolished in March 1917 and already in November 1917 the Bolsheviks came to power. So the traditions of 'authoritarian, Byzantine type' rule and of disregard of the law are strong and those of defending individual rights and making one's demands known are not yet sufficiently developed.

There has never really been a sustained tradition of "service for the public" and observing the principle of impartiality in the Russian state administration. The tendency towards behavioral conservatism in various organs of public administration is deeply rooted and democratic traditions are virtually absent. During Soviet times such fundamental principles of state administration as ethics, transparency and accountability were either unknown or understood quite differently to the way they are perceived in modern civil services.

The existing administrative culture regards secrecy as a virtue, whereas responsiveness, transparency and dialogue with citizens are generally viewed as matters of secondary importance at best.

Another significant feature of the administrative culture in Russia is rather low respect for law. A representative survey conducted by the Russian Academy of Civil Service among federal civil servants in 1999 showed that in performing their duties near 70% of civil servants gave priority to following the instructions of their superiors and only second priority to compliance with existing rules and regulations (Grishkovets, 2002).

Russian society has changed dramatically during the last decade. Many of the above mentioned values and traditions appear to have been affected in recent years by new "standards" coming from the West and regretfully not always the best ones. Even so, the legacy of the past continues to be a powerful factor exerting a strong impact on administrative developments in Russia as shown in the subsequent chapters of this study.

3.16 Accountability and Responsibility

Key principles of any democracy is the requirement that the bureaucracy should bear responsibility, must be subject to scrutiny, and made accountable to legislature, state controlling agencies, the electorate and civil society at large.

The demise of the communist system created a situation when the former "master" - the communist party, which exercised strict control over conduct of its members and citizens - disappeared. The replacement system and democratic institutions were not sufficiently strong or mature to ensure that the bureaucracy effectively serves the democratic state and the people. As often happens, the reality has fallen well short of the ideal.

The very first decree of President V. Putin, when he assumed office in 2000, is remarkable in many ways. This decree granted the former President B.Yeltsin immunity from any type of criminal or administrative persecution. This legal act provides good illustration of the status of accountability of some state officials in Russia.

Several points should be stressed here:

(i) The Law on State Civil Service adopted in 2004 provides for several types of responsibility of civil servants. They are subject to disciplinary, administrative, criminal and material (compensation of damage) responsibility. However, in reality this responsibility is almost exclusively internal;

(ii) The recent decade saw a steady erosion of accountability and responsibility accepted by the Government and federal ministers. “Decisions in the executive branch in essence are taken collectively, nobody carries personal responsibility”. (Otechestvennyie Zapiski, 2004:86);

(iii) Furthermore, within prevailing clientele/patronage networks responsibility and accountability of civil servants to the society and citizens have been replaced by accountability to those who authorises appointments and promotions.

"The administration was never accountable to the general public," Mikhail Dmitriev, the first Deputy Minister for economic development, said in an interview to the "International Herald Tribune" on 19 July 2006. "It was never in fact responsible for delivering a high quality service to society. Ordinary citizens were considered subjects to be ruled who themselves should serve the state."

During most of the 1990s and 2000s the President and the Government could largely control decision-making in the Parliament and the latter’s role of counter-weight in the system of "checks and balances" was seriously undermined. As a result accountability of the executive to the legislature has been weak. For example, the decision of the Parliament to carry out an audit of the activities of the government apparatus in 1994 was effectively blocked by the Prime-Minister. In another case, in January 2003 the deputies of the Duma were infuriated by the incapacity of the Government to cope with chronic failures of heating systems, which left over 50 thousand people in the Russian North defenseless against severe frost. The deputies requested the Prime Minister to come and report on the situation to the Parliament. The Head of the Government refused and send his Deputy instead.

3.17 Ethical Standards and Corruption

The latest decade has witnessed a considerable decline of moral and ethical standards in society and in the civil service. The excessive involvement of the state in economic activities, growth of mixed forms of public/private ownership, reduced oversight and control over the bureaucracy by the legislature and civil society exposed civil servants to wide opportunities for obtaining personal gain.

Since the late 1980s corruption has grown considerably in scale and variety of forms it takes and at present poses a serious threat to the democratic gains and advance of reforms in Russia. Apart from reaching new proportions it has become generally less risky, more cynical, cash oriented and profitable. In view of many experts corruption in Russia has become virtually “institutionalized” to the extent that there exists a market of state portfolio and services. Many civil servants are involved in commercial activities and in the 1990s some commercial entities have been defining state policy in certain areas.

The notion of “conflict of interest” is only being introduced within the framework of the civil service reform and even now instances of its formal exposure are rare. Often civil servants do not realize that their conduct is improper. Such a situation is profitable for large businesses which benefit significantly from protectionist practices and government patronage.

Estimates made by experts of "Technologies – XXI century" show that businesses in Russia spend minimum USD 500 million a month (USD 6 billion a year) on bribing federal and regional officials. A poll conducted by the Foundation "Public Opinion" in four regions of Russia in 2004 revealed that over 76% of citizens encountered cases of corruptive behavior of civil servants. (Yuzhakov, 2005).

Already back in 1992 President B. Yeltsin signed a decree "On fighting corruption in the system of the civil service". Failure to enforce its basic provisions became evident very soon. From the outset it was envisaged that the decree would be followed by a respective law. However, almost ten years of fight to get various anti-corruption draft laws passed through the Parliament has so far brought no result. Since 1998 several bills on fighting corruption were submitted by a group of deputies to the Duma, each time to be rejected by the Duma committees. Another draft was tabled in 2007 and is still in the Parliament.

In his annual address to the Federal Assembly in 2006 President V. Putin made the following statement "despite all the efforts we have made, we have still not yet managed to remove one of the greatest obstacles facing our development, that of corruption."

The Presidential Administration and the Government have recently established an Anti-Corruption Board to elaborate and coordinate a programme of anti-corruption measures. However, the precise mission and authority of the Board has not been defined yet. Available information indicates that the Board had only two meetings during the two years following its formation.

Many observers express scepticism over the prospects of steady advance of the anti-corruption campaign prior to the presidential elections in 2008.

3.18 Assessment of Available Resources

Successful operation of any civil service is contingent, among other things, upon adequate funding and availability of well educated and motivated professional staff. Throughout the 1990s the Russian civil service was operating under severe budget constraints. One of the implications was low remuneration which could no longer serve as a serious motivating factor for the majority of civil servants. At that time many government agencies experienced frequent shortages of even very basic items such as paper or funds for intercity telephone communications.

In 2000 only 10% of interviewed civil servants were quite confident in the stability of their employment, 33% were somewhat confident and 47% pointed to their lack of confidence in this respect (RACS, 2002:187). As we have stressed above in the 1990s many professionals left the civil service for more lucrative jobs in the private sector. As a result, by the early 2000s there had grown serious deficit of specialists of up to 45 years of age knowledgeable in and capable of applying modern management technologies (Federal Programme 'Reform of the Civil Service of the Russian Federation', 2002). In his address to the Federal Assembly on the 16th of May 2003, President V.Putin remarked that 'despite a huge number of state officials… this country finds itself in dire need of modern managers.'

In the 1990s the percentage of the federal state budget expenditure allocated to finance the system of public administration in Russia amounted on average to 3% (RACS, 2002:148). In the majority of industrialized countries the respective figure is on average 8-10%. In 2002 this percentage grew to 3.59% (Savchenko, 2006:19) and continued to increase in the subsequent years. However, in the view of some experts the official data do not reflect the real expenditure of the state in this area and suggest an under estimate by a factor of two (Nikolaev, 2006).

Deficit of proper motivation of civil servants, in particular absence of direct linkage between pay and performance, is another factor which undermines the effectiveness of the Russian civil service and its potential for reform. As a result, for example, the state resources are either wasted at best, or used for enriching corrupt officials at worst.

3.19 Civil Service and Civil Society

One of the peculiarities of the Russian transition to market economy and democratic society consisted in the absence of a long term and steady interaction between the "new" political elite and the society at large as well as lack of a single uniform system of values accepted both by the elite and masses.

The formation and development of the new political institutions in contemporary Russia was not so much the result of "bottom up" pressure from the society but rather emerged as the product of the "top down" policies pursued by relatively narrow but dynamic post-communist elites. This also contributed to widespread alienation of citizens to the political institutions imposed from above.

The civil society, which could have played a positive role in stimulating adherence to democratic principles and strict accountability across all levels of government, is still at an early stage of development. The widespread political apathy and abstention of the population created favourable conditions for the subordination of the public administration to narrowly based interests.

The idea of a bureaucracy both responsive and accountable to the people has not yet penetrated the "social consciousness". Instead, the state apparatus is often viewed as a "thing in itself", as a corps of professionals who in their activities are guided above all by the "corporate" interests of the bureaucracy and by wishes of the "powers that be". The latest surveys show that 72% of civil servants reckon that their work should be assessed as good whereas 71% of citizens are not satisfied with civil servants' activities (Nezavisimaya Gazeta, 2006:10).

A poll conducted among Russian citizens and civil servants in 2001 to establish their assessment of moral qualities of civil servants gave the results shown in Table 3.2.

Table 3.2 Assessment of the moral qualities of the Russian civil servants

	Opinion of civil servants	Opinion of citizens
Majority of civil servants are honest, incorrupt and committed	22.8%	13%
Number of honest and dishonest civil servants is approx. equal	58.6%	51%
Majority of civil servants are not quite honest and corruption resistant and few are committed	8.9%	30%

Source: the Russian Civil Service: Analysis of Formation and Personnel Development, RACS, 2002, Moscow, p.326

The following arguments are most often used by citizens in explaining their rather negative perception of civil servants: a) majority of civil servants do not know real needs of the people; b) they enjoy special privileges although at the beginning were against such preferences themselves; c) they are preoccupied only with the fight for their place near the "feeding trough"; and d) they do not seek opinion of citizens. Among other factors that undermine positive image of civil servants in society are: lack of transparency, the gap between words and deeds, and the instances of corruptive behavior (RACS, 2004 b: 28-31).

It would be incorrect to conclude that most civil servants in Russia are not dutiful and law abiding employees. However, prevailing in the Russian civil service, particularly in the 1990's, instability and patronage practices, complemented until recently by highly unattractive remuneration conditions, have seriously distorted civil servants' traditional incentives - patriotism, loyalty and sense of duty - and devaluated such merits as responsibility, honesty and respect for the law.

The majority of the civil service cadre is not ready for performing their duties under the conditions of transparency and accountability and to treat citizens as clients. Opinion polls carried out in the 1990s among civil servants showed that on average only 10% of the respondents singled out "greater oversight from citizens" as one of the methods of removing shortcomings and deficiencies in the civil service. (RACS, 2002 a:14).

A survey of the heads of personnel and training departments of the federal and regional executive agencies (autumn 1997) demonstrated that only 10% of those interviewed considered strengthening the public relations component as one of priority topics for training of civil servants. Analysis of the statutes of PR departments of regional administrations revealed that only 30% of them are tasked "to ensure transparency and openness of the activities of the administration" as one of the priority activities of the departments (RACS, 2002a:16).

The deeply embedded culture of secrecy and treatment of citizens as subordinates is not easy to overcome. Certain measures have been introduced in recent years (though far from sufficient) to secure opportunities for ordinary citizen to gain some form of redress in situations of unfair treatment by authorities or material damage inflicted by the latter's actions. Steps have also been taken to improve citizens' access to the information about the administrative regulations and decision making. These include, inter alia, establishment of the Ombudsman office, approval of legislation that allows citizens to claim compensation for the damage inflicted by the state bodies and civil servants, and submission to the Duma of the draft Law 'On Free Access to Information About Government Activities'.

3.20 Conclusions

Since the 16th century the Russian state administration has been developing within the political and cultural environment characterized by either strong elements of absolutism and despotism or non-democratic form of governance. The main orientation of the state administrators has always been to the state and primarily to the ruler of the state. The state bureaucracy enjoyed a privileged status and was actually exempt from effective social control. Its most typical features were strong authoritarian tendencies, neglect of the needs of common people, and opposition to the society at large. The traditions of 'authoritarian, Byzantine type' rule and of structural and behavioural conservatism were deeply rooted and commitment to democratic principles was virtually absent.

The situation in this respect did not change much under the Soviet system. The following bureaucratic characteristics were typical of the soviet state administration: rigid and complex hierarchy with decision making function only at the top of the command structure; authority derived from personal ties rather than merit; and prevalence of rigid rules undermined by vertically organized patronage.

The Soviet Russia system of governance crumbled in the early 1990s when the new political leaders committed themselves to democratic principles, the economy was opened to the market and the Soviet Union ceased to exist. By 1991 the system of nomenklatura in Russia was largerly demolished.

The early 1990s were marked by some positive developments in the Russian public administration. These included adoption of a basic regulatory framework aimed at the formation of a modern civil service as a distinct legal and social institution, a set of specialized activities and functions, and unity of professional cadres with relevant resources.

However, it appears that measures undertaken throughout the 1990s to modernize the civil service were episodic, inconsistent, uncoordinated, and with no sound strategy geared to perspective. The biggest drawback of the adopted legislation was its weak enforcement.

In spite of the profound transformation of the political and economic system in the 1990s, many outdated structures and procedures within the state administration as well as inherited from the Soviet times incentive and motivation schemes have been left largely intact. A merit system has been declared a fundamental principle of the civil service but in practice has often been substituted by personal loyalty. Authoritarian tendencies and elements of clannish bureaucracy remained deeply rooted.

The Russian bureaucracy managed to retain its influence, in fact strengthened it further by accumulating state property and resources and additionally reducing control to which the bureaucracy had been previously subject. Opportunities for promoting "corporate" interests of the bureaucratic establishment have eventually increased in scope.

A prominent Russian economist made the following statement: "During recent 15 years of economic and political reforms, in spite of all turbulence and uncertainty, there has been only one steady and constantly advancing process in Russia, - the process of liberation of state bureaucracy from any control by the society and from any accountability to the society" (Delyagin, 2005). This statement may be regarded as exaggerated to a certain extent but is true in principle. A recent OECD report described Russia in a similar way as a "weak state with strong bureaucrats" (OECD, 2006:1).

By the mid 1990s Russia had largely adopted a career based model of civil service. Many elements of this model were borrowed from the French civil service system. This can be explained by both similarity of the Russian and French Constitutional orders, with a strong Presidency being one of their distinguishing common features, and by close collaboration between Russian and French officials and experts on public administration in the development of conceptual approaches and documents (Zaitseva, 2003:69). Contrary to the French model, the Russian model retained many outdated practices, rules and traditions. In practice the distinction is manifested above all in considerably lower extent of accountability of the bureaucracy to the parliament and to the citizens than in France.

Throughout the decade that followed the demise of the communist rule the capacity for civil service reform in the Russian Federation remained overall low. This capacity was undermined by the following major factors:

- excessive politicisation of the civil service;
- domination of informal patronage networks;
- widespread corruption;
- absence throughout the 1990s of committed reform stakeholders, e.g. mature civil society, influential business sector, strong political opposition;
- shortage of qualified expertise;
- severe financial constrains (until early 2000s);
- lack of political will to launch civil service reform until the election of the new President in 2000.

The reform capacity was further affected by unpredictability and uncertainty of the political and economic situation in Russia and overall absence of a clear and sound long term strategy for the country's development.

Summing up, formation in modern Russia of an efficient, accountable and corruption immune civil service turned out to be an extremely complicated task and in spite of certain achievements by the early 2000s the bulk of the effort was still lying ahead.

CHAPTER IV
REFORM OF THE RUSSIAN CIVIL SERVICE: WHO WILL WIN THIS TIME – BUREAUCRACY OR POLITICIANS?

4.1 Introduction

The previous chapter was devoted to the study of the status of the Russian civil service and its capacity for reform. Consistent with the logic of our research we now turn to the analysis of the actual preparation and implementation of civil service reform in the Russian Federation.

Following our methodological approach we start with review of numerous attempts to modernize the state apparatus in the post-communist Russia during the 1990s. Then we examine the reform driving forces and the circumstances, both favourable and unfavourable, for its initiation and implementation.

The reform's conceptual documents, its overall dimensions and directions are then analysed. This analysis is complemented by study of the reform related legislation and the processes of the administrative reform launched almost simultaneously with that of the civil service.

Subsequently we explore the reform's preliminary outcomes, as well as difficulties and constraints revealed during the process of its implementation.

Through the course of the analysis special attention is paid to a set of issues relevant to the nature of civil service reform as defined in the Protocol for comparative studies of national civil service systems. These issues comprise, in particular, the reform's major stakeholders, its objectives and implementation mechanisms, its results and the extent of external influence on the selection of the reform's agenda and priorities.

Positive and negative experience in implementing the reform of the Russian civil service is summarized in the conclusions to this chapter.

4.2 Pressure for Change

Systems of civil service are not static and evolve to reflect the social and economic pressures as well as the interplay of different political forces. By the late 1990s it became clear to many observers that the outdated and practically unchanged paradigm of the Russian state administration not only failed to meet the new challenges but was actually in conflict with the requirements of Russia's prospective development. In his 2005 annual address to the RF Federal Assembly President V. Putin emphasized that "Our bureaucratic apparatus is still largely an exclusive and often arrogant caste viewing state service as an alternative form of business. Therefore, our priority remains making state management more effective, ensuring that officials strictly obey the law and quality public services are provided to the population".

The Russian Minister of Economic Development and Trade G.Gref stated in July 2002 "Russia is losing up to 2% of GDP because of poor public management". Even more impressive are the losses caused by corruption in the state administration, estimated by various sources to be in access of USD 35 billion annually (INDEM, 2003).

Public opinion polls conducted in 1999 showed that over half of respondents assessed negatively the work of Russian civil servants. Of those interviewed 66% attributed their negative attitude to widespread corruption in state administration, 48% to red tape and arrogance, and 48% to neglect of the interests of citizens (RACS, 2002:11). Recent surveys point to further deterioration of these tendencies.

Throughout most of the 1990s, in the absence of a clearly defined institutional framework for decision-making and authority, reform minded politicians and officials both in the Russian Government and the State Duma (Parliament) tried independently to develop initiatives aimed at modernization of state administration. The political and financial crisis in 1998, growing fragmentation of the state, upsurge of corruption, criminalization of the economy, steep decline of public's confidence in the President and Government all made implementation of in-depth civil service and administrative reforms an even greater and more urgent priority than it was in the early 1990s.

For many years there has been also lack of unity of political and social forces that could apply effective pressure to eventually achieve preparation and launch of civil service reform. Their capacity to influence political decision-making process was weak. Equally, civil society in Russia has not yet reached a sufficient state of maturity and therefore was not in a position to serve as a reform inducing factor.

All this has led to the emergence of various contradictions. For example, one could note the contradiction between civil servants who were oriented to advance in their career by means of high professionalism and those who were relying mainly on the mechanism of protection and personal loyalty; contradiction between free market operators who expected fair competition and its inhibition by numerous administrative barriers artificially and selectively created by the bureaucrats; or contradiction between the declared objectives and tasks of the state administration and lack of clear division of authority among government agencies complemented by insufficient resources for their achievement.

By the late 1990s considerable pressure for initiation of reform came from the distinct negative tendencies in the economic and social development of the country, worsening of practically all parameters of performance of the government and state agencies and lowering of living standards of the majority of the population. These tendencies could not have been disregarded by the politicians who wished to count on any respect and support of citizens.

4.3 Preparation of Reform

In Russia, where the state formerly performed all basic functions (many of which in a modern society are often executed by independent organisations and associations), the very notion of civil service and the role it should play within the context of new democratic institutions and market oriented economy have been subject of intensive debate and controversy. Following the break up of the Soviet Union in 1991 and subsequent collapse of the government infrastructure the new political leadership of Russia declared transformation of the communist state apparatus into an efficient state administration capable of promoting and regulating market relations and acting in a new democratic environment as one of its top priorities. However, during the 1990s the reform of state administration lagged far behind political and economic reforms in the country.

It should be noted that the first commission on the elaboration of a draft law on civil service was established in the Soviet Parliament (Supreme Soviet) in 1989 i.e. already during the advanced stage of perestroika and prior to the ban on the communist party and break up of the Soviet Union. By 1991 the commission prepared the first draft Law "On the Federal Civil Service of the USSR". That was an absolute innovation in the history of the Soviet state though one of whole series of new developments in various spheres of social and political life at the later stages of perestroika. The draft law stipulated the formation of the Commission of the Civil Service of the USSR as its managing and control body. This initiative did not materialize at that time though it laid the foundation for subsequent plans to develop civil service legislation. Nonetheless, in 1992-1993 the Russian Parliament continued its work on the preparation of draft civil service legislation. The political crisis in 1993 put this process on hold.

Following the formation of the Russian Federation as a sovereign state in 1991 the issue of transformation of the state administration and control over it became highly politicized. Different lobby groups attempted to push their schemes of the organisation and reform of the state administrative apparatus. At least three separate drafts of civil service reform were circulating during the period 1992-1995, each reflecting different interests of particular groups of players.

An attempt was made in 1991 to set up a special central organ to manage state administration - Department on civil service issues under the President of the Russian Federation. However, due to a number of political and organizational constraints these plans failed. Already at that time the importance of full political support for the reform from the highest levels of state and government became quite evident.

In 1992, thanks to the efforts of the reformers, a new structural unit 'The Chief Department on Personnel Training' (*Roskadry*) was created under the Government of the Russian Federation. This Department was made responsible not only for training of government officials, but also for developing concepts of reforming state administration and respective legislation. Activities of *Roskadry* contributed to the adoption in 1993 of the first normative acts concerning the civil service. They were the Presidential decrees "On Priority Measures to Organize a System of Civil Service in the Russian Federation" and "Regulations on the Federal Civil Service" signed by the President respectively on 3 June and 22 December 1993 (Atamanchuk, 2002:63). These documents outlined general provisions with regard to the organization and operation of the civil service at the federal and regional levels.

Simultaneously experts of Roskadry worked on the elaboration of the concept of civil service reform. The guidelines of draft concept of reform of the Russian civil service were published for the first time in the Rossiyskaya Gazeta on 23-24th December 1993. The concept was based on the assumption that legal, normative institutional establishment and development of the civil service will take at least several years and in view of overall political and economic uncertainty in the country the approach to reform should take a form of small steps rather than "big bang". The contents of the document reflected largely the vision of a career based, state oriented civil service.

Regretfully from its very origin *Roskadry* lacked a clear-cut mandate and authority to ensure appropriate policy making capacity. Declining political support to its mission led to abolishment of *Roskadry* in early 1994.

In February 1994 the Department of Federal Civil Service of the President of the RF was established (Zaitseva, ed., 2003:284). This was followed in June 1994 by the creation of the Russian Academy of Civil Service – a central national educational institution responsible for training of middle and top level civil servants.

Following many years of controversy and debate, the Law "On the Basic Principles of the Civil Service of the Russian Federation" was finally adopted by the Parliament in summer 1995 (three years after the publication of the first draft). It became the first basic legal document which laid the foundation for the proper organization and functioning of the Russian civil service. The Law's basic provisions stipulated the creation of a career civil service system.

In the mid 1990s a debate on the future of the Russian civil service led to the emergence of two groups of experts each advocating essentially different approaches to the civil service and its reform. The first approach (often called institutional) was based on the assumption that civil service is a social and legal institution serving above all interests of citizens, society and the state. The second approach (called political) viewed civil service primarily as professional activities ensuring execution of authority of state bodies and realization of the policies mapped out by the political forces in power.

By the mid 1990s Russia had largely adopted a model based on the second approach. In elaborating essential elements of this model experience of the French civil service was widely used. This could be explained both by similarity of the Russian and French Constitutional arrangements and by close cooperation of the Russian and French officials and experts on public administration in drafting the conceptual approaches and documents for the development of the Russian state administration (Zaitseva, ed. 2003:69).

In the late 1990s elaboration of civil service reform concepts in Russia was largely conducted as part of preparation of a broader administrative reform. A prominent role was played by a group of the closest advisors to President Yeltsin. They tried to push the President for the start of in depth transformation of the system of governance and civil service. This time the developers of the reform concept advocated creation of the model based on the first (institutional) approach.

In March 1998 President Yeltsin made an announcement about the forthcoming administrative reform that would drastically transform the whole branch of executive power in Russia. He promised to make public the concept of the reform drafted by a team of his advisors and assistants. Although the text of this document was never published (which came as no surprise for the *connoisseurs* of bureaucratic behaviour), the available information about its contents allowed conclusion that it was in many ways a remarkable document.

For the first time the programme of administrative reform (including reform of the civil service) was based on a comprehensive strategic approach that established priority issues and a sequence of actions in implementing reform. It comprised some rather 'revolutionary' (at least for Russia) measures. The concept envisaged, for instance, abolishment of the powerful Apparatus of the Government and considerable downsizing of the President's Administration, as well as delegation of greater authority and responsibility to the ministries, the number of which was to be also reduced. The concept was generally in line with the modern trends in administrative and civil service reform. It addressed such important and burning issues of civil service development in Russia as performance based pay, merit based evaluation and career advance, impartiality and accountability, fight against corruption, provision of quality services to citizens, enhancement of inter-agency co-operation, improving relations between federal centre and regions, etc. The document stated that the transformation of the administrative apparatus would take around 8-10 years. However, as often happened to reform initiatives in Russia, in particular in the sphere of governance, these measures remained mostly on paper.

"Even if there were no other reasons for the failure to implement the reform programme of 1998, there were no sufficient political desire and attention to support it. Although the programme was officially approved and acknowledged, the Yeltsin administration lacked sufficient understanding, stamina or willingness to implement it." (Larjavaara, 2001:56).

By the early 2000s the changes introduced into the Russian administrative system did not appear to have followed any specific pattern, were adhoc and largely cosmetic. These changes were not designed to alter the nature of the existing administrative system and were far from being comprehensive. Boris Yeltsin did little to change the essence of the previous administrative system and modernize the state apparatus because in this respect the prevailing situation suited him. He used the outdated structures and mechanisms to consolidate his position by acting as arbitrator in bureaucratic battles stemming from the poorly demarcated division of authority.

4.4 Reform Driving Forces

It is widely acknowledged (as we have shown in Chapter II) that political will and political leadership are the crucial factors in any far reaching changes affecting the administrative institutions, structures, styles and culture.

In spite of President Yeltsin's occasional lavish rhetoric regarding the need for civil service reform it was never followed by equally strong action. For example, in his state of the nation address in March 1997 President Yeltsin emphasised that it was time to restore order above all in the governing bodies at all levels: 'The authorities are getting fat..., they are only concerned with their own well-being... There is corruption at every level of power ... The Government, not circumstance, must lead the country". He promised a fundamental reform of the executive power but gave no details how he intended to carry it out. The 'reform' itself was later confined to a limited set of measures, mainly liquidation of several ministries and appointment of new Deputies to the Prime Minister. True, there existed (as we have shown above) a small group of advisors and experts which tried to initiate reform but their policy making ability turned out to be limited. By 2000 the mandate of President Yeltsin to reform the civil service has been largely unfulfilled.

Society was in clear need for a more consistent and objective approach regarding reform of the country's numerous bureaucracy and this fact seems to have been clearly realized by the new political leadership, namely, President V.Putin. On 24 November 2000, just six months after being elected, the new Russian President sent a letter to the Prime Minister M. Kasyanov requesting that proposals for the civil service reform be prepared by 1 May 2001 (Zaitseva, ed. 2003:172).

Under President Putin preparation for reform was accelerated with the issue in December 2000 of a directive by the Presidential Administration on establishment of an inter-agency commission. This body was charged with drafting the Concept of Civil Service Reform and subsequent establishment of three subgroups to deal with reforms of the military service, law enforcement agencies and the civil service respectively.

The President's address to the Federal Assembly in April 2001 gave an important impulse to this work. The President pointed out that development of the huge potential of the country is blocked by the cumbersome, inert and inefficient government machinery, while officials who master modern management methods are few and far between. The President acknowledged that the officialdom resisted reforms, and that the current system of operation of executive and legislative authorities, oriented largely to extracting the so-called "status rent", was dangerous to society and the state.

Much of the conceptual and analytical work on the civil service and administrative reform was done at that time in the Gref's Center for Strategic Studies (CSS) - one of the leading Russian think tanks (G.Gref was Minister of Economy from 2000 till 2007 – A.K.). This work was carried out within the framework of the efforts to elaborate the country's strategy of economic and social development till 2010. By summer 2000 experts of the Gref's Center submitted a concept of civil service reform comprising a tentative action plan (Zaitseva, ed., 2003:209). This concept made up a large part of the special section of the above mentioned strategy devoted to the issues of proper governance and public management.

The authors of this concept advocated preparation and launching of a series of institutional reforms covering: budgeting; public administration (mainly government re-structuring); civil service; deregulation; judicial matters; law enforcement agencies; military; and relations between the federal centre and subjects of the federation. The Administration of the President was supposed to play a leading role in managing the reform agenda.

In August 2001 a presidential edict was issued on establishment of the Commission on Civil Service Reform headed by Prime Minister M. Kasyanov. In the same month the President approved the Concept and elaboration of the Program of Civil Service Reform got underway. Simultaneously, a legal act that would amend and improve the law 'On the Basic Principles of the Civil Service in the Russian Federation' (adopted in 1995) was being drafted.

On 19 November 2002 the President signed the Presidential decree "On the Programme of Reform of the Russian Civil Service (2003-2005)" giving the "green light" to the actual implementation of the reform which started in early 2003.

It is noteworthy that M. Dmitriev, First Deputy Minister of Economic Development and Trade remarked in one of his interviews in 2002 that three years ago it was senseless to put forward issues of civil service reform for consideration by the Government. "We simply would not have been heard" he stated.

Simon Kordonsky, one of the President's senior advisors said in an interview that "the main client and stakeholder of the administrative reform is the President". When questioned whether there were forces inside the state apparatus which supported and pushed for administrative reform and if there was a team of reformers within the ruling elite he replied "I have not heard about such team" (Otechestvennyie Zapiski, 2004:94).

President Putin appears to have committed himself to reform the state administration but, as available evidence shows, he himself was not sure if his efforts would be successful. Celebrating his one year anniversary in office President Putin told a group of journalists that the main lesson he had learnt was "the extreme difficulty of fighting bureaucracy" (The Observer, 8 June 2003). The embedded conservatism and non-acceptance of the reform ideas by the bureaucratic establishment was clearly demonstrated when the above mentioned document 'Strategy of Economic and Social Development of the Russian Federation till 2010' was published in the mass media. It transpired that the section on improving the existing system of governance was the only one missing in the text.

Overall one may say that internal imperatives for reform stemmed from the apprehension amongst farseeing representatives in government, political and economic circles and large businesses that preservation of the largely unreformed and outdated state machinery will lead Russia into consolidation of its status as a raw materials supplier at best and to collapse of its political, economic and social fabric at worst. The Russian President was one of those who appear to have realized this to the fullest extent.

4.5 Reform Preparation and Implementation Stages

The conducted analysis allows single out the following stages in the preparation and implementation of Russian civil service reform:

I. STAGES OF REFORM PREPARATION 1992–2002
a) 1992 – 1995: during this period efforts were undertaken to develop a normative framework (Presidential decrees, Government resolutions, Federal laws) for the formal establishment of the civil service and transformation of the existing state administration culminating in the adoption of the Law 'On the Basic Principles of the Russian Civil Service' (1995);
b) 1996 – 1998: in this period various concepts of civil service and administrative reform were elaborated by several groups of experts;
c) 1999 – 2002: concept and programme of civil service reform elaborated by Gref's Center of Strategic Studies and later further amended and improved by the Inter-Agency working group, were finally approved by the President in August 2001 and November 2002 respectively.

II. STAGES OF REFORM IMPLEMENTATION 2003 – until present
a) 2003 – 2005: the first stage of the Reform programme based on a federal targeted programme approach was implemented;
b) 2006 – 2007: implementation of the programme of civil service reform continued further to its extension to 1 January 2008;
c) 2008 – 2013: implementation until 2013 envisaged in the draft Programme of reform and development of the Russian civil service elaborated by the Presidential Administration.

4.6 Civil Service Law from 1995

Adoption by the State Duma in July 1995 of the Law "On the Basic Principles of the Civil Service of the Russian Federation" was a prime landmark in the modern history of evolution and transformation of the Russian state administration and deserves special attention. In spite of some obvious deficiencies, this legal act gave for the first time in the modern history of Russia an official definition of the civil service, laid down the legal basis for its organisation, set a distinction between purely political and administrative appointments, outlined the procedures of service and introduced a Job Classification System.

Elaboration of this law was accompanied by intensive political debate on what kind of civil service should exist in Russia and who should be granted direct authority over it. This indicates clearly the significance of the issue for major political actors. The controversy over the preparation and adoption of the law had an impact on its contents which combines modern and traditional (outdated in views of many observers) requirements for the civil service mission, status, activities, organization and procedures.

Among the undoubtedly positive provisions introduced by the Law were formal requirements for holding qualification examinations, regular performance review, competitions to fill a vacancy, regular submission by civil servants of income declarations, and conforming to restrictions imposed on civil servants. The Law fixed the principle of impartiality and non-partisanship of civil service. Consistent and strict application of these requirements would have provided for the establishment of a "merit based" civil service in Russia.

Regretfully, full enforcement of these provisions contained in the Law turned out to be problematic. This can be explained by both the overall weak institutional environment for law enforcement in Russia and by resistance of the bureaucracy and various lobbies whose interests were affected by the Law.

On the other hand, the Law virtually restored an outdated and complicated bureaucratic "table of grades and ranks". It also left unchanged (no effective mechanism envisaged against it) the system of almost total dependence of a civil servant on his/her superior who has discretion to decide the issues of career advance, level of take home pay and various granted privileges. Thus, the law did not remove fertile ground for the continuation of the "clientelistic" and "patronage" practices prevalent in the Russian civil service.

The Law neither clearly defined the place of the civil service in the context of institutional roles and division of powers, nor differentiated clearly the authority and functions of civil services at the federal and regional levels. It did not provide a mechanism for the protection of the civil service from arbitrary political interference and the establishment of a single management body for the whole civil service. Although the Law reserved a place for trade unions within the system, collective bargaining was not envisaged. The Law retained reference to the Labour Code - "civil servants shall be subject to the labour legislation of the Russian Federation with account for the specific provisions hereof" (Article 4, point 3 of the Law).

In essence these elements in the Law oriented development of the civil service towards consolidation of its "ruler dependent" and "patrimonial" characteristics.

4.7 Concept of Civil Service Reform

International experience demonstrates that a uniform vision of the directions and contents of a civil service reform among its major stakeholders is a rather rare phenomenon and Russia is not an exception in this respect. However, by the late 1990s there has built up at least an understanding among major political forces in Russia that unreformed state bureaucracy constitutes a serious obstacle in the path of the country's economic and social development. In the opinion of many experts the planned transformation of the bureaucratic apparatus was to result in the creation of a professional, accountable and impartial civil service based on merit and immune to corruption.

One of the key priorities in Russia at that time was reform of the entire system of public management. This was considered important because reform of the civil service without appropriate transformation of the entire system of governance and public management does not make much sense. Indeed, real improvement of the efficiency of civil servants' work and enhancement of their responsibility can hardly be achieved whilst there remains several decision-making centres working in parallel within the executive branch. Similarly, it is difficult to expect that the civil service would stand up to modern challenges if the issue of clear division of authority between the federal and regional executive bodies remains unresolved. Therefore, from the outset civil service reform was conceived as part of a wider administrative reform.

As we have shown above the concept of civil service reform was developed in response to the letter signed by President V. Putin on 24 November 2000 requesting that proposals on reforming the system of the Russian civil service be submitted to him by 1 May 2001. The contents of the concept approved by the President on 15 August 2001 was largely based on the workings and contributions made by the authors of Administrative reform concept (1998) and experts of the Gref's Center for Strategic Studies who developed a draft concept of civil service reform in summer 2000.

The concept's basic idea was to gradually transform the Russian state administration into a legitimate and efficient institution serving above all the society and the state. The concept outlined the main objective of the reform which consisted in raising effectiveness of the civil service and creating an integral holistic system of civil service in the Russian Federation. The contents of this document reflected sensitive points of the status of the Russian civil service and proposed solutions to many longstanding problems. The stepwise approach chosen by the authors of the concept to be implemented during a lengthy period seemed to be rational. The concept outlined the following principal tasks of the reform:

- adjustment of the civil service system and civil servants' professional activities to the new social and economic reality;
- development of comprehensive civil service legislation;
- introduction of effective mechanisms of HR management;
- formation of the system of management of civil service and coordination of its activities;
- ensuring openness and transparency of civil servants' activities;
- enhancing quality of services rendered to citizens;
- introduction of modern information technologies;

The Concept devoted much attention to the establishment of an integral and uniform system of civil service; regulation of civil servants' activities on the basis of job descriptions; introduction of standards for delivery of services to people; interaction of civil service with civil society; removal of conditions conducive to corruption; improving staff training programmes, etc.

Preparation of the reform concept as well as of other reform documents has been largely nontransparent and in this respect was no difference to the previous attempts to reform the civil service. For a long time the public was not given any information about who was preparing the reform, how this was being done, and what reform was all about. Four months elapsed between approval of the concept by the President and publication of its very general description. The full text of the concept was made available to the mass media almost a year after approval by the President (Grishkovets, 2002:126).

4.8 Programme of Civil Service Reform

The elaborated and approved concept of civil service reform served as the basis for the development of the reform programme. The latter was designed by members of the interagency working group on civil service reform issues set up on 13 December 2000 (Savchenko, ed. 2006:77). On 19 November 2002 the President of the Russian Federation approved the Federal Programme 'Reform of the Civil Service of the Russian Federation for the period 2003-2005'.

Apart from the analysis of the existing situation and identification of the reform objectives and tasks the programme comprised a rather general action plan for the period 2003-2005. The plan outlined measures aimed at resolving the defined tasks, achieving overall programme objectives and keeping up with the time table for the reform implementation. The Russian Government committed to allocate 539 million rubles (equivalent of approximately USD 17.4 million) for implementation of the Programme over 3 years (see detailed description of the Programme in Annex 1).

The principal goal of the programme (as specified in section 2) was – *improving efficiency of the civil service as a whole and of its structural elements, optimizing civil service maintenance costs and developing resources required for civil service activities.*

To achieve this goal, the programme (section 2) identified the following objectives:

a) Creating conditions for optimal organizational and legal support of the civil service;
b) Defining roles, powers, and responsibilities of civil servants based on job descriptions;
c) Implementing new methods and techniques of planning, financing, stimulating, and assessing civil servants' activities, and using civil service system resources in a rational manner;
d) Ensuring openness of the civil service for the benefit of civil society development and strengthening of the state;
e) Applying efficient methods for selecting skilled personnel for the civil service and for assessing performance of civil servants, as well as creating conditions for civil servant career advance;

f) Implementing staff training programs for the civil service and professional development programs for civil servants;
g) Implementing mechanisms for identifying and solving civil service-related conflicts of interest, as well as introducing a legal regulation of professional ethics of civil servants;
h) Creating optimal material and technical conditions for efficient functioning of the federal civil service and performance by civil servants of their official (service) duties;
i) Ensuring the development of civil service management system.

The programme paid particular attention to the application of result-oriented management; introduction of standards for state services and administrative regulations; ensuring openness and transparency of state institutions; rebuilding the incentives structure for civil servants; development of partnerships with civil society; recruitment and retaining of highly qualified and professional personnel; development of an efficient system of training of civil servants; enforcing effective anti-corruption policy.

The so called federal targeted programme (FTP) was declared the key instrument of reform implementation. FTP is a result oriented mechanism of carrying out large scale programmes of social, economic and institutional development, which envisages close coordination of all stakeholders and agencies, detailed planning and funding from various sources including federal and regional governments.

The programme (section 6) defined the following anticipated results:

- Harmonization of the regulatory framework of the civil service with the existing social relations and new economic environment;
- Establishment of a professional merit based civil service;
- Introduction of modern job descriptions for civil servants;
- Achievement of high-quality performance by civil servants of their official (service) duties and provision by them of public services to citizens and organizations;
- Creation of conditions for openness and accountability of state bodies' administrations and civil servants to the civil society;
- Creation of a framework for improving allocation of financial and economic resources to the federal civil service;
- Improvement of the efficiency of HR policy within the civil service system in order to improve the quality of civil service employees structure;
- Implementation of mechanisms for identifying and resolving civil service related conflicts of interest, as well as introduction of legal regulation of professional ethics of civil servants;
- Elevation of the professional level of federal civil servants;
- Optimization of the number of federal civil servants;
- Creation of a material and technical environment for efficient performance by civil servants of their official (service) duties;
- Development of a new civil service management system.

In the opinion of the Head of Civil Service Directorate of the RF Presidential Administration Mr. Savchenko realization of the Federal Programme should eventually provide for: a) creation of an integrated civil service system in the

Russian Federation; b) considerable increase of the efficiency of activities of civil service; c) formation of highly skillful and qualified personnel in the civil service; and d) creation of a modern system of civil service management. (Savchenko, ed., 2006:45).

The programme envisaged that before full-scale implementation is attempted. pilot projects would be run in selected ministries and regions. In practical terms the stages of the realization of the programme (2003-2005) foresaw the following succession of major activities: a) adoption of the respective legislation and normative base, including *the Law 'On the Civil Service System of the Russian Federation'* and *the Law 'On the State Civil Service of the Russian Federation'*; b) development of the methodological, organizational and financial mechanisms of the programme implementation; c) implementation of pilots and experiments in various government institutions and regions; and d) broad implementation of the reform ideas, principles and provisions across the entire civil service (Savchenko, ed., 2006:39).

More specifically the programme envisaged implementation of pilot projects targeted at:

- approbation of new methods of planning of activities of federal government organs;
- elaboration of criteria and indicators of assessment of activities of civil servants;
- application of result-oriented principles of funding of selected state organs;
- introduction of service contracts including fixed term;
- testing new systems of career advance, professional training and up-grading;
- introduction of effective mechanisms of identification and prevention of conflict of interests.

As we can see the programme incorporated objectives and tasks which fall largely in line with modern trends in civil service reform practiced world wide. However, the programme deliverables did not appear sufficiently specific which made assessment of the progress in its implementation rather difficult. Furthermore, an attachment to the programme pointed only to a particular year when a result or measure should be achieved: this cannot be considered as a proper deadline. Partly this could be explained by a prevailing diversity of opinions concerning reform contents. Deputy Minister of Economic Development and Trade M. Dmitriev stated in May 2002 that "consensus regarding this reform agenda was superficial" (Dmitriev, 2002).

4.9 Reform Preparation and Implementation Milestones

November 1991 - Decision of the RF President (Presidential decree N 242) to set up the Main Directorate on Civil Servants Training (Roskadry) under the Government of the Russian Federation; this structure was made responsible for civil servants training and elaboration of conceptual documents related to civil service reform;

March 1992 - The Presidium of the Supreme Soviet (Russian Parliament) adopted enactment N 2535/1 "On the draft law regarding civil service of the Russian Federation";

December 1993 - President signed a decree (N 2267) "On the Regulations of the Federal Civil Service";

June 1994 - Establishment of the Russian Academy of Civil Service by the Presidential Decree N 1140;

July 1995 - Adoption of the Federal Law N 119 "On the Basic Principles of the Civil Service of the Russian Federation";

October 1998 - Draft Concept of administrative reform developed by the Presidential advisors;

July 2000 - Draft Concept of civil service reform elaborated by Gref's Center of strategic studies;

November 2000 - President V. Putin signed a letter *(poruchenie)* to the Prime Minister requesting that proposals for the civil service reform be prepared by 1 May 2001;

December 2000 - Inter agency working group on civil service reform issues set up by resolution of the Presidential Administration N 1331;

June 2001 - Commission of the RF Government on reducing administrative barriers in entrepreneurship and optimization of budget expenditures in public management (headed by the Minister of Finance A. Kudrin) established by the Government resolution N 452;

August 2001 - Approval by the President of the Concept of Reform of the Russian Civil Service (Decree N 1496);

August 2001 - Creation by Presidential Resolution N 436 of the Commission on Civil Service Reform headed by the Prime Minister;

October 2001 - Establishment of Inter agency working group on elaboration of the Programme of civil service reform and respective legislation (Resolution of the Presidential Administration N 1331);

August 2002 - Approval of the Code of Ethics of civil servants by the Presidential decree N 855 "On General Principles of Professional Conduct of Civil Servants";

November 2002 - Federal Programme "Reform of the Civil Service of the Russian Federation (for the period 2003-2005)" approved by the Presidential Decree N 1336;

May 2003 - Federal Law (N 58) "On the Civil Service System of the Russian Federation" adopted by the State Duma;

July 2003 - The President signed a Decree N 824 "On Measures to Implement Administrative Reform During the Period 2003 – 2004";

November 2003 - Anti-Corruption Council under the President of the Russian Federation set up by the Presidential Decree N 1384;

February 2004 - The Department of Federal Civil Service was established in the Presidential Administration by the Presidential Decree N 298;

March 2004 - Large scale reorganization of the Russian Government enacted by the Presidential Decree N 314;

July 2004 - The Presidential Decree N 910 "On Measures to Improve State Governance" established a Commission under the RF President for Improvement of State Governance headed by the Head of the Presidential Administration. The Commission comprised three Working Groups: (i) Interagency Working Group on Implementation of Civil Service Reform Measures; (ii) Interagency Working Group on Federative Relations and Local Self-Governance; and (iii) Interagency Working Group on the System and Structure of Federal Bodies of Executive Power;

July 2004 - Federal Law (N 79) "On the State Civil Service of the Russian Federation" approved by the Duma;

October 2005 - Adoption of the Concept of administrative reform and Action Plan for the period 2003 – 2004;

December 2005 - Extension by the Presidential Decree N 1437 of the Federal Programme "Reform of the Civil Service of the Russian Federation" till 2008;

December 2005 - Decree N 1574 "On Register of the RF State Civil Service Positions" signed by the RF President;

July 2006 - Decree N 763 "On Monetary Compensation for Federal Civil Servants" signed by the RF President;

December 2006 - Decree N 1474 "On Post Entry Training of the Russian Civil Servants" signed by the RF President;

December 2006 - Decree N 1490 "On Increasing Monthly Salaries of Federal Civil Servants" signed by the RF President;

March 2007 - Federal Law N 25 "On the Municipal Service of the Russian Federation" signed by the RF President;

March 2007 - Decree N 269 "On the Establishment of the Commissions on the Observation of the Requirements for the Professional Conduct of State Civil Servants" signed by the RF President;

June 2007 - Decree N 722 "On Conducting Experiments within the Frameworks of the Reform of the Russian Civil Service" signed by the RF President;

December 2007 - Draft Federal Programme "Reform and Development of the Russian Civil Service (2008-2013)" elaborated by the Presidential Administration.

4.10 Reform Implementation and Management

Since the start of the reform implementation in early 2003 the Administration of the President of the Russian Federation has performed the function of coordinator of the reform activities. It was also assigned the task of monitoring and controlling reform implementation. "The Program Coordinator shall be responsible for organizing the Program implementation and monitoring progress in carrying out the measures envisaged therein" (Federal Programme "Reform of the Russian Civil Service", 2002). The structural unit in the Presidential Administration in charge of reform coordination and monitoring has been the Directorate on Civil Service Issues.

The principal authority to drive civil service reform at the nationwide level first rested with the Commission on Civil Service Reform headed by the Prime Minister (established in August 2001) and later transferred to the Commission for Improvement of State Governance (established in July 2004) headed by the Head of the Presidential Administration. The Commission had the status of a consultative body to the President of the Russian Federation and was supposed to have three inter-agency groups: a) On Civil Service Reform; b) On Federal Relations; and c) On the Issues of the System and Structure of Federal Executive Agencies.

The Inter-Agency Group on civil service reform was in charge of developing conceptual approaches and adopting key decisions on this issue. It consisted of officials representing all branches of power and comprised an affiliated permanent team of approximately 200 experts. The Inter-Agency Group was led by Chief Presidential Advisor (Savchenko, 2006: 18).

Meetings of the above commissions were held quarterly on average. In essence the commissions constituted ad-hoc consultative/coordinating structures with restricted mandate (especially after July 2004) to manage reform and make important decisions obligating all stakeholders. In reality a special body directly responsible for managing and effectively exercising reform has not been established. There has been weak coordination with other reforms, above all with administrative reform.

Reform management was particularly weak at the operational level. Insufficient attention was paid to monitoring and assessment of progress as well as timely resolution of the emerging problems and correction of the reform tactics and approaches. This prevented reform managers to address operational issues in an effective manner.

The reform management was further compounded by rivalry between various groups of bureaucrats for power and authority within the state administration itself and definite shortage of relevant technical expertise. Deficiencies were evident in particular of experts with practical experience in modern human resource management, result oriented budgeting and prevention of conflict-of-interest situations. Capacity of various regions to absorb and implement reform measures was uneven (Otechestvennyie Zapiski, 2004:109).

The reform programme referred to the following methods and means of achieving reform objectives:

a) selection and formation of groups of experts and specialists;
b) development and approval of methodologies, selection of pilot regions and ministries;
c) implementation of pilots and experiments in the selected regions and government agencies;
d) enhancing training, retraining, and professional development of federal public servants;
e) improving the legislative and regulatory framework for civil service;
f) enforcing the provisions of the regulatory and legal framework in practice.

Section 5 of the Programme stipulated that: "The Program shall be implemented using federal budget funds on the basis of state contracts (agreements) concluded according to the established procedure by state customers with implementers of the measures envisaged by the Programme and on the basis of agreements concluded with relevant constituent entities of the Russian Federation or municipalities".

The state contracts were grouped according to the reform tasks. The basic idea was that successful implementation of the contracts should facilitate fulfilment of the programme tasks and create favourable ground for the achievement of its results at least those timed for short term perspective. Therefore particular consideration was given to linking contractual activities as much as possible to the planned results. This approach makes up an important element of the federal targeted programme (FTP - referred above) or sometimes called *target oriented method of implementing reform* (Savchenko, 2006:14).

Out of 539 million roubles allocated from the Federal budget for reform purposes 296 million roubles have been geared to research and development work. The remaining 243 million roubles were used to implement pilot experiments and other activities.

Analysis of available information on reform progress and opinions of the experts and civil servants involved in reform implementation indicates that most reform activities during the first three year period (2003 – 2005) were concentrated on: a) elaboration of civil service legislation and normative framework; and b) preparation and implementation of pilots in selected ministries and regions.

During the period from 2000 to mid 2004 overall 117 laws and decrees of the President and Government resolutions were adopted on various issues of the civil service and its reform (Otechestvennyie Zapiski, 2004:112). During two years that followed adoption of the Law 'On State Civil Service of the Russian Federation' in 2004 more than 20 Presidential decrees were issued on various aspects of civil service regulation.

Considerable funds have been allocated and overall consistent efforts have been made so far to implement various pilot projects and experiments (both at the federal and regional levels) aimed at raising the efficiency and transparency of civil servants' activities. Some of the pilots were carried out within the frameworks of the on-going administrative reform and projects funded from external sources (mainly the European Commission, DFID, and the World Bank). As a rule, the launch of pilot projects was preceded by the elaboration of appropriate methodology for their implementation. Active implementation of pilot experiments was concentrated mainly in five-six Subjects of the Federation (regions).

Pilots and experiments covered a wide range of issues including in particular – *development and application of performance indicators for individual civil servants; elaboration and application of standard employment contracts; elaboration and application of job descriptions for various categories of civil servants; elaboration of normative acts to provide for establishment of regional civil service management structures; and improving the effectiveness of the use of reserve cadre.*

In January 2005 President Putin signed a decree extending for two years (until 2008) the programme of civil service reform. The decree provided for continuous flow of funding for the programme implementation whereas its goals and tasks remained in principle unchanged.

4.11 New Civil Service Legislation

The civil service reform programme (2003-2005) put high emphasis on the development of modern legislation and regulatory framework for the Russian civil service. In this respect particular importance was attached to the adoption of two federal laws: the Federal Law N55 of 27.05.2003 "On State Service System of the Russian Federation" and the Federal Law N79 of 27 July 2004 "On the State Civil Service of the Russian Federation".

The first Law "On State Service System of the Russian Federation" stipulated that there should exist three types of state service in Russia: the civil service, the law enforcement service and the military service. The civil service consists of two levels: the federal civil service and the civil service of Subjects of the Russian Federation. Status compatibility across the three types of state service was established. Thus every civil service position corresponds to a certain position in the law enforcement service and similarly within the military service. The law also envisaged establishment of the body in charge of management of the service at federal and regional levels.

The second Law "On the State Civil Service of the Russian Federation" was originally developed to replace the civil service law adopted in 1995. Therefore, this legal document set out basic legal, organisational and economic principles of the Russian civil service. It also defined the rights and duties of civil servants, determined conditions and terms of service and career advance and fixed appropriate requirements and restrictions.

Principal innovations and changes contained in the new laws as compared to the law from 1995 were as follows:

- four new categories of civil servant positions were introduced;
- qualification classes were replaced with civil service ranks;
- reference to Labour Code as governing legislation for civil servants was entirely dropped;
- fixed term contracts with a wider range of compensation linked to performance in distinction to the standard rules of compensation became possible;

- conflict of interest regulation was introduced;.
- frequency of compulsory re-training of civil servants was increased from every five years in the previous law to every three years;
- establishment of a single management body for the federal civil service and similar bodies in each Subject of the Federation (region) was envisaged;
- up to 25% of the members of the competition commissions were to be independent experts.

The above mentioned innovations were overall positively assessed by the majority of experts and observers. However, the new legislation also included clauses and provisions that raised controversy and evoked criticism. In particular, it was alleged that the legislation lacked certain principles and requirements that were expected to be part of the new normative environment.

These include failure of the new legislation to introduce essentially new reward mechanism. The latter remains largely based on the principle of seniority and length of service. Performance based pay is limited almost exclusively to fixed term positions in the civil service which make up a tiny portion of the overall number of contracts.

This means that career advance of a civil servant will continue to depend predominantly on the individual's relations with the relevant chief and less on professionalism, competency, skills and efficiency of work.

The laws did not provide for the creation of effective mechanism opening civil service to public oversight and scrutiny.

Although the Law of 2004 stipulates that recruitment and promotion should be competition based (Ch. 4, art. 22) there are numerous exceptions to this rule. For example, exempt from this requirement are civil servants with fixed term contract, those with access to classified information, those appointed by the President etc. Competition is also not mandatory in cases when position can be filled by persons from the so called cadre reserve (Ch 4, art 22). Taking into account that such candidates are available in the majority of promotion cases, this clause considerably undermines the practical value of the principle of competition. According to the Law a list of cadre reserve nominees also should be compiled on a competitive basis. However, in practice this rule is seldom observed and many experts express doubt as to the objectivity of competition in such cases.

It is noteworthy that the number of clauses in the new law granting exceptions from the requirement to hold mandatory competition as far as enrolment and promotion are concerned has actually increased compared to the law from 1995.

The new laws extended the age limit for holding civil service post to 65 years whereas before the ceiling was fixed at 60 years of age.

It appears that the requirement of non-partisanship of civil servants and separation of religious communities from the state organs (Article 5 point 11 in the Law of 1995) was dropped in the new legislation. No mechanism to effectively prevent excessive politicisation of the civil service has been foreseen.

The new laws failed to create a uniform and integrated civil service. Regional authorities retained considerable independence in determining the specific organisation of the regional civil service.

In general the new laws appear to have largely preserved the existing situation rather than responded to a new one. It is noteworthy that both laws on civil service "were prepared in total secrecy without involvement of public organizations and citizens" (Yuzhakov, 2005). An interesting definition was given to the new laws on civil service (adopted respectively in 2003 and 2004) by State Duma Deputy V.Yuzhakov – "both laws are laws about bureaucrats, drafted by bureaucrats and in the interests of bureaucrats" (ibid).

4.12 Administrative Reform

It is widely acknowledged that civil services can function properly only within the framework of well designed administrative structures and processes with clearly defined roles of executive units, adequate horizontal and vertical coordination systems, clear lines of accountability and transparent policy making mechanisms.

Therefore, civil service reforms are often made part of wider administrative reforms or are accompanied by comprehensive reform measures in the administrative environment.

As we have indicated at the beginning of this chapter the first concepts of administrative reform were developed in the Russian Federation as early as 1997-1998. Administrative reform was officially launched in July 2003 by the Presidential Decree N 824 "On Measures to Implement the Administrative Reform in 2003-2004". In the same month the Government set up the Commission on Implementing Administrative Reform headed by the Deputy Prime Minister.

The decree of the President 824/2003 formulated the main objective of the administrative reform – "*to raise effectiveness of the public administration in particular in providing services for citizens, enhancing the rule of law, removing administrative barriers in dealing with the business sector*".

The same decree defined the following measures aimed at improving efficiency of the activities of the government and creating a more favourable climate for business:

- reduction of government intervention into activities of businesses, including abolition of excessive government regulation;
- elimination of duplication of the functions and powers of the government agencies;
- creation of a new structure of government with clearly separated functions of policy making, regulation, supervision and control; of managing government property; and of providing government services to citizens and legal entities;
- delineation of powers and authority between federal and regional levels of government.

In the years 2003-2005 funding of administrative reform from the federal budget amounted to USD 7 million each year (Otechestvennyie Zapiski, 2004:179).

In spring 2004 within the framework of the administrative reform a comprehensive restructuring and reorganisation of the Federal Government was launched which turned out to be much more radical than most observers predicted. In fact this was the largest reorganization of the Russian government over the last 40 years.

A three category system of federal executive bodies has been introduced comprising ministries ("setting the rules"), services ("enforcing the rules") and agencies ("implementing the rules").

- Ministries were made responsible for the delivery of Government Program objectives within their area of competence. The main preoccupation of Ministries is to *"set the rules"*, undertaking policy analysis and drafting legislation;
- Services were entrusted with *"enforcing the rules"* through supervision and regulatory activities;
- Agencies were assigned the task of *"implementing the rules"* through service delivery and management of state property.

The number of ministries has been reduced from 24 to 15. The number of Deputy Prime Ministers was cut from six to one. By the Prime Minister's decision 146 government commissions, advisory committees and groups have been dissolved. Regular weekly meetings of the Cabinet were opened to the press during fixed periods of time. However, the total number of executive bodies at the federal level has increased from 53 to 72.

Apart from a number of key regulatory bodies, such as the Federal Financial Markets Supervision Service and the Federal Anti-monopoly Service which report to the Prime Minister, all Services and Agencies were subordinated to a specific Ministry. Simultaneously staff reductions were announced at the level of ministries. For example, within the Ministry of Economic Development and Trade the number of departments has been cut from 58 to 16, the number of Deputy Ministers from 17 to 2 and total staff by 300.

Such a radical restructuring process could not fail to increase uncertainty and insecurity among the majority of civil servants and in most cases led to a significant reduction in their performance.

In accordance with the introduction of the three categories system of executive organs a review of the functions of federal government bodies (functional reviews) was conducted in 2003-2004. The Government Commission headed by Deputy Prime Minister screened over 5600 functions. Out of this number 800 were found excessive, 350 duplicating and 500 requiring reduction of the scope of application (World Bank, 2005:110).

Implementation of administrative reform in the regions proceeded at a slower rate with only 21 Subjects of the Federation (regions) adopting a programme of administrative reform by 2005.

One of the peculiarities of the administrative reform in Russia was that it was launched prior to the elaboration of the reform's concept. A Presidential advisor admitted at that time that there did not exist a document where the goals, tasks and mechanism of administrative reform were outlined (Otechestvennyie Zapiski, 2004:86).

The concept of administrative reform and action plan for the years 2006-2008 were developed and approved later by the Government Resolution N 1789 of 25 October 2005. These documents defined the following priority directions of reform activities:

- development and introduction of service standards;
- changeover from funding of government agencies based on their budget estimates to allocation of funds for specific programmes to be implemented;
- enhancement of senior executives' rights to redistribute funds between budget items and to use the payroll at their discretion;
- wider usage of information technologies;

- increasing transparency and accountability of government structures.

The reform implementation mechanism was modified to allow for a more decentralized approach where each ministry or Subject of Federation developed their own programme based on methodological support provided centrally. Also a competition based principle of funding reform implementation from the state budget was introduced. The 2006 state budget allocated an equivalent of approximately 18 million USD for this purpose.

4.13 Impact of External Diffusion and Assistance

As we have stressed above the objectives, directions and methods of the reform of the Russian civil service were largely in line with modern trends in designing and implementing administrative reforms practiced world wide.

There are no firm grounds to assert that any specific foreign concept of civil service reform was borrowed as a model for replication in Russia. It is known that in 1997-1998 the Russian experts drew upon the experience of the United States ('National Performance Review'), France (civil service legislation) and Germany (post-war administrative transformation) in elaborating the concept of administrative reform.

In the late 1990s and early 2000s the scale of international cooperation and donor assistance in the area of public administration expanded and became further diversified.

However, we should be aware that by the mid 1990s Russia has largely adopted a career based model of civil service, many elements of which were borrowed from the French civil service system. As we pointed out earlier this can be explained by similarity of the Russian and French Constitutional orders with a strong Presidency being one of their distinguished features. There was also close collaboration of the Russian and French officials and experts on public administration in the development of civil service legislation and reform's conceptual documents throughout the 1990s (Zaitseva, 2003:69).

Contrary to the situation in the Czech Republic where the declared process of accession to EU entailed a considerable pressure from Brussels on the Czech authorities to implement genuine civil service reform, there was no such outside pressure experienced by the Russian authorities. The reference to the necessity of public administration reform in World Bank and IMF key documents relating to Russia started to be clearly articulated only in the years 1999 – 2000 but were not really binding for the Russian leadership.

Various international organizations and separate countries have been delivering technical and other assistance to Russia in support of the development of its public administration and civil service reform. The European Commission has been the largest player in the field earmarking around 323 mln euros over the period 1991 - 2003 in support of institutional, legal and administrative reform in Russia (Tacis Action programme, 2004:130).

The experience of Russia shows that international assistance to public sector management and civil service reform has been so far focused mainly on achieving short-term results without considering longer term consequences. External assistance could have been more systematic and better coordinated. In connection with this one observer made the following comment: "The international intervention programmes in Russian civil service reform have accordingly tended to be either piecemeal, or 'pilot' in nature, and therefore uncoordinated, and hard to 'scale up', thus generating what is in effect a self-fulfilling prophecy about the country's 'openness to reform'. Some external interventions have by contrast been excessively 'foundational', assuming that development must start from a long way 'back up the modernity ladder', and thus perversely have the effect of reinforcing Russian 'backwardness' with legislative measures which have little chance of being implemented or are at such a high level of generality that they fail to address the pressing operational difficulties faced by Russian government day to day". (Gray, 2005:14).

Establishment in January 2004 in Moscow of the Donor Secretariat funded by DFID (UK) to coordinate assistance to public administration reform offered by external donors appears to have contributed to increasing efficacy of the foreign intervention and ensuring its integrity with the Russian Government priorities.

4.14 Assessment of Civil Service Reform Progress and Mid-term Results

As we noted above assessment of the progress of civil service reform in Russia is rather difficult because of absence of clearly defined outputs and specific deadlines in the reform programme as well as frequent shortage of data about the extent of advance of reform.

Judging by the available information, interviews with participants and rare statements made by those who were directly involved in coordinating and managing reform activities we may state that notable advance of civil service reform has been registered mainly in the following areas:

a) adoption of legislation and other normative acts (laws, decrees, enactments);
b) considerable increase in monetary compensation of civil servants;
c) implementation of pilots and experiments in selected ministries and regions.

Apart from the two laws "On the State Service System of the Russian Federation" and "On the State Civil Service of the Russian Federation" adopted in May 2003 and July 2004 respectively over 30 legal and regulatory acts were signed by the President and the Head of Government by mid 2006 as part of the secondary legislation specifying and complementing the provisions of these laws. These acts dealt with such issues as arrangements for resolving conflict of interest, organising competitive recruitment, awarding class ranks, carrying out appraisal of civil servants, conducting qualification examination, approval of the roster of civil service positions and others.

In connection with this a World Bank report noted "For the past two and a half years significant progress has been achieved in developing legal and methodological framework for the reform implementation" (World Bank, 2006).

Implementation of the reform Programme together with creation of material and technical base for efficient functioning of the civil service has brought about some quite tangible results. It is estimated that during the period 2004–2007 the monetary compensation of Russian federal civil servants increased on average 100-

150% for mid level positions and 300-500% for top level positions. The Federal State Statistics Service (Rosstat) data published in March 2007 revealed that in 2006 average remuneration of federal civil service employees increased by 24% and amounted to 13,425 roubles which was higher than general Russia average of 10,728 roubles. In some Russian regions civil servants appear to be in a particularly privileged position in this respect. The same Rosstat data demonstrated that in 2006 the average salary of a civil servant in St. Petersburg was 30,990 roubles (approx. USD 1200) which was twice higher than the average salary in the city and higher than average civil servant's salary in Moscow ("Kommersant Daily", 27 March 2007).

The advance in implementing pilots and experiments, though not even, was overall successful as it laid down the ground for practical enforcement of the innovative approaches contained in the reform Programme and legislation, provided there was sufficient will for these changes.

Some of the experiments appear to be truly innovative in the Russian context. For instance, in line with the Presidential Decree signed late 2006, selected federal ministries and agencies were allowed to retain the pay bill of those staff members, who have left (for whatever reason) public office or were dismissed, and to use the savings made to increase the salaries of the best performing civil servants (Sharov, 2007).

In another case experimental introduction of performance related pay took place in 2006 in 20 government agencies and offices rendering services to citizens and businesses. The experiment envisaged considerable bonuses to the staff on top of their standard remuneration schemes in cases where the customers were satisfied with the quality of rendered services (Sharov, 2007).

Activities were carried out along other directions of reform as they were mapped out in the Programme but with limited scope and overall less convincing results. Starting from 2005 considerably more state budget funds were directed at training civil servants. This allowed to increase the number of staff trained and to diversify the forms of training, e. g. to use more long term and distance learning courses, to send more civil servants for training abroad. However, the shortcomings of the existing system of training - very formal competition among training institutions (often organised only for subsequent reporting); low fees for the teaching staff; and in most cases weak relevance of training programmes to the real practical needs of civil servants - basically remained.

Similarly, introduction of modern job descriptions in the majority of cases turned out to be formal. The Presidential Decree N 159/2005 "On the Official Contract of Employment by the Civil Service and Appointment to Civil Service Positions" set three months period for the development of job regulations by all federal government institutions in line with new requirements. However since no methodological guidelines had been developed to help ministries in drafting new job regulations the latter transformed the existing job instructions into job regulations without any substantive changes (World Bank, 2006:25).

In spite of much rhetoric a number of key issues in the reform programme, in particular, excessive politicisation of the civil service and fight against corruption have been left unattended in terms of achievement of any practical results.

The task of further democratization of civil service implying introduction of mechanisms for increased accountability to democratically elected institutions and civil society, transparency and openness, client orientation and responsiveness, consultation with citizens appears to have been so far of "second priority". For example, in his annual address to the Federal Assembly in 2003 the President stressed the necessity of developing effective mechanism for resolution of disputes between a citizen and state through improving administrative and judicial procedures. However, a year after this statement very little if anything had been done in this respect. (Otechestvennyie Zapiski, 2004:114).

The managers of the reform programme appear to have defined development of civil service legislation as the principal reform implementation tool. This decision does not seem to be fully justified for a number of reasons. First, because of weak law enforcement in Russia in general and of civil service related legislation in particular. Secondly, because there has been definite shortage of qualified and skilful experts in Russia capable of drafting legislation and other normative documents on civil service issues. Thirdly, because this priority overshadowed other important implementation tools such as networking, coalition building and fiscal incentive instruments which turned out to be partially or entirely "forgotten".

The Law "On Procurement of Services and Goods for the State Needs" provides good illustration of weak law enforcement. The Law envisaged that almost all purchases should be tender/competition based whereas in reality nine out of ten procurement procedures were exempt from competition (Dmitriev, 2005:42). Another example is linked to the Law "On the State Civil Service of the Russian Federation" which envisaged measures to regulate conflict of interest situations. As a result, commissions on conflict of interest were set up in the majority of ministries and other state institutions. However, in course of the two years following adoption of the law in 2004 no single case of such a conflict was reviewed (ibid).

One of the weaknesses of reform implementation was evident shortage of committed reform stakeholders. Mobilisation of internal and external support for reform has been largely disregarded. As a result only 11% of interviewed civil servants were ready to fully support the reform, 79% were ready to do it only partially, and 8.5% were not supportive of reform at all (RACS, 2004:17).

Reform implementation has been accompanied by a number of unforeseen side effects. It turns out that while Russia's population has decreased steadily in recent years, the number of civil servants has increased at an inversely proportionate rate.

According to the data of the Federal State Statistics Service only in 2005 the number of state administration employees at the federal level increased by 10.9%. A report by the consulting company Troyka Dialogue indicates that in the course of three years (2004-2006) the number of civil servants had grown by approximately 17%, whereas labour productivity in the public sector had dropped 7% for the same period.

Even more interesting is that funding of public administration from the state budget for the period 2000–2006 increased 11 times in nominal prices whereas, for instance, expenditures on social protection grew 3 fold only. (Business Week, Russia, 28 August 2006). The same period saw expansion of privileges for top level civil servants. For example, since 2005 officials from the level of Deputy Head of Department upwards gained the right for business class travel by air and 1 class travel by train.

Summing up we may conclude that the civil service reform has failed so far to make considerable advance in most of priority directions as mapped out in its programme. That means that the announced transfer to an effective, merit based and accountable civil service is still pending. One of the top managers of civil service reform, Head of the Commission on Improvement of State Governance, D.Medvedev remarked that "the reform got stuck in bureaucratic problems, in inability to organize work and activities in an appropriate manner" (Izvestia, 27 December 2005). Similar opinion was expressed by the World Bank report, which noted that civil service reform "has so far not moved significantly, overall from the status quo" (World Bank, 2006:7).

4.15 Assessment of Administrative Reform Progress

Implementation of the administrative reform in Russia was taking place in parallel to civil service reform. Many observers expressed the view that the issues of integrating approaches to both reforms and of coordinating their advance do not appear to have been properly taken care of.

Most radical in recent Russian history, the restructuring of the Russian Government that took place in spring 2004 was a widely acknowledged current result of the administrative reform. However, the assessment of this restructuring has not been straightforward. Having liquidated 24 ministries, committees and agencies the Government created 42 new state institutions where most of the dismissed officials found new employment. Since that time the number of officials has been constantly growing (Izvestia, 11 November 2006). The reform reduced the number of Deputies to Prime Minister from 6 to 1, and Deputies of Ministers from between 6 and 15 (depending on the ministry) to only 2 in each ministry. But by 2007 the Prime Minister had again 4 Deputies, and Federal Ministers raised the number of their Deputies to 4-5 on average.

Restructuring led to the reduction of the staff of the Office of the Government by 17 % (2,100 persons) and territorial branches of the Government agencies cut a further 42,000 persons (Izvestia, 20 April 2005). However, as we have shown above, in less than a year the overall number of civil servants actually increased as compared to the pre-reform period.

In the end government restructuring turned out to be limited to rather mechanical mergers of government institutions without thorough consideration of the level of preparedness of these structures for the reform. In one of the interviews given late 2004 First Deputy Minister of Economic Development M. Dmitriev gave the following assessment of the measures undertaken within the scope of administrative reform; "The crisis of executive branch manageability is obvious. It is a typical transitional crisis, where reforms are incomplete and inconsistent. The functions of executive branch bodies have been inventoried, the government structure revised, and a three level administration system including services and agencies introduced. This kind of structure has a number of advantages as it helps to quickly take decisions and makes the administrative system more transparent. On one condition, however, namely if the structure changes are supplemented with new administrative methods such as changes in the principles and standards of civil servant's work, the introduction of a new result driven remuneration system, and fight against corruption. Too little, if at all, has been done in this regard" (Dmitriev, 2004).

Certain tangible results were achieved in removing administrative barriers. In particular, entrepreneurs were granted the right to open business in five days subsequent to submission of a prescribed number of legal documents for registration. The number of control and check up procedures to which businesses had been previously subject was considerably reduced by introducing fixed frequency and advance planning. The overall number of business activities that needed to be licensed was reduced from 214 to 120 (Otechestvennyie Zapiski, 2004:107).

Also tested under the administrative reform were new approaches and mechanisms for planning, budgeting, usage of resources, motivation and performance assessment. The reform programme envisaged that the regional authorities would adopt similar reform programmes funded both from the federal and local budgets. By early 2006 such programmes had been approved in less than half of the Russian regions.

To conclude, in spite of certain setbacks in reform implementation, civil service and administrative reforms seem to remain important issues in the government agenda in Russia. Extension by the President of the Programme of civil service reform to 2008 and approval in October 2005 of the Concept and Action plan of the administrative reform for the period 2006–2008, as well as submission by the Presidential Administration in late 2007 of the draft Programme "Reform and Development of the Russian Civil Service" for the period 2008-2013 provide additional evidence for that statement.

4.16 Conclusions

In 2002 Russia committed itself to radical changes in its system of state administration by approving a programme of reform of the Russian civil service. The task of transforming the administrative apparatus which the Russian Federation inherited from the old Soviet system into an efficient, merit based, impartial, responsive, corruption immune and accountable state administration turned out to be clearly an immense one.

Although many progressively thinking politicians and top ranking civil servants realized already in the early 1990s that the growing problems faced by the nation could not be successfully solved with unreformed state administration, for many years this apprehension was not turned into practical action. Throughout the whole 90s political will to launch reform of the Russian civil service was absent. Like many other major reforms in the history of Russia the impetus to the reform of the civil service came from the top. The reform was initiated by newly elected President V. Putin who has been its principal driving force.

From the very beginning civil service reform had been conceived as part of a wider agenda, above all part of the reform of system of governance. However, in reality the extent of integration with other reforms like administrative reform, reform of judiciary and budgetary reform was in many cases superficial.

Raising efficiency of civil servants' activities was declared the priority objective of civil service reform. To achieve this purpose the Programme envisaged such measures as introduction of performance based pay, improvement of post entry training, development of modern job descriptions, setting up a system of civil service management, regulation of situations of conflict of interest and others. Development and introduction of new regulatory framework transpired to be the principal tool of reform implementation.

The major goals of the administrative reform (apart from the reorganisation of the government) which was launched almost simultaneously with civil service reform consisted in the removal of administrative barriers in dealing with the business sector, improving standards of services provided to citizens and introduction of performance related budgeting.

Compared to the period of stagnation of the Russian civil service throughout most of the 1990s recent changes induced by the reform constitute a significant shift for the better. In terms of meeting the declared objectives of the civil service and administrative reforms the achieved results appear so far rather modest. They are confined mainly to the adoption of civil service related legal and normative framework, implementation of pilots and experiments in different regions of Russia, increase in civil servants' pay, wider scale of training (civil service reform), restructuring of the government and removal of some administrative barriers in dealing with businesses (administrative reform).

The authors of principal conceptual reform documents drew widely upon the experience of other countries in designing and implementing civil service reform. It would be wrong to assert that any specific foreign concept of civil service reform was borrowed as a model for replication in Russia. Even so, for a number of reasons the influence of representatives of the French civil service has been significant especially during the 1990s.

Implementation of civil service reform revealed certain drawbacks and weaknesses that constrained considerably its advance and achievement of key results.

The process of reform preparation and to a certain extent implementation has been largely non-transparent. Potential stakeholders (e.g. civil society, business community) were not invited to take part in designing conceptual documents and monitoring reform progress. Mobilisation of internal and external support for reform was largely disregarded.

There turned out to be no real ownership of reform. The majority of civil servants were either indifferent to reform at best or remained in silent opposition at worst.

Some issues important for the development of the Russian civil service, such as excessive politicization and embedded corruption, have not been properly addressed yet.

High priority given to legislation as the main implementation tool turned out to be detrimental to wide application of other tools and instruments. As a result such tools as best practice diffusion and training were used rather inconsistently. Networking and fiscal incentive instruments were largely ignored.

Advance of reform was further affected by insufficient operational management, weak law enforcement and proven ability of the bureaucracy to distort or sidetrack many of the innovative approaches contained in the reform programme.

Finally, the analysis of the reform revealed certain unforeseen side effects. These include considerable increase in budget spending on state administration and notable growth of the number of civil servants at both federal and regional levels in recent years.

One of the principal conclusions that can be drawn after several years since the start of reform is that it has failed so far to change the internal incentive mechanism within the civil service and introduce effective systems of external accountability. Career advance of Russian civil servants continues to depend more on their relationship with the respective higher level management and much less on their performance.

CHAPTER V
THE CZECH CIVIL SERVICE: HISTORICAL LEGACY, CURRENT SITUATION AND CAPACITY FOR REFORM

5.1 Introduction

The objective of this chapter is to explore reform capacity of the Czech civil service. To achieve this objective and in line with our methodological approach the chapter starts with review of the historical development of the Czech nationhood and evolution of the state administrative apparatus on the territory of the present day Czech Republic. Particular focus is made on the prevailing traditions and stereotypes of behavior of state administrators. Further, we examine the overall institutional framework of the Czech civil service and in particular, peculiarities of the Czech constitutional order and political system, and economic development of the country since the early 1990s including the process of privatization. Our attention is also focused on the civil service's legal status, political-administrative relations, issues of bureaucratic accountability and ethical standards.

Subsequently societal and administrative traditions, values and culture as well as relations civil service–civil society are analyzed. This analysis is complemented by the study of the selected features of the Czech civil service system: overall structure and human resources management, civil servants' incentive mechanism and training, professionalism of cadres and available financial resources, and problems in contemporary development of the service.

It is important to bear in mind that from a formal point of view the Czech civil service had not existed until 2002 as there was no respective law or other legal act providing for its legal recognition. However, state administration in the Czech Republic has been functioning in a normal mode and possessed many features common to those inherent to bureaucracies in most industrial countries. Hence, in spite of temporary absence "de-jure" of the subject of research, it has existed "de-facto" and therefore could be respectively analyzed and compared.

Most information in this chapter relates to the processes taking place in the Czech public administration and its environment throughout the 1990s prior to the adoption of civil service law. Overview of recent developments will be given in the next chapter.

5.2 The Czech State and Its Administration - Historical Overview

An independent Czechoslovak state was established only in 1918 but its roots go back many centuries. The history of the Czech nation is predominantly a history of co-existence and struggle with powerful neighbors.[1]

---

[1] Source: http://www.geocities.com/TheTropics/Resort/3353/history/index.html
http://www.czech.cz/en/czech-republic/history/all-about-czech-history/
http://www.mongabay.com/reference/country_studies/czech-republic/history.html

The Moravian Empire

Around the 4th century B.C. the present-day Czech Republic was populated by Celts. They were the first ethnic group to arrive in the area, according to available historical evidence. The Celtic Boii tribe gave the country its Latin name - Boiohaemum (Bohemia). They were overrun by the German tribes before the beginning of our era.

Between the 5th and the 6th centuries, during the period known as the Migration of Peoples, the tribes of Slavs moved to the territory of Bohemia and Moravia and settled there. The first half of the 7th century marks the first successful attempt to unite Slavonic tribes. The so-called "Samo's kingdom" resisted the pressure of the powerful Avar empire centered in the Hungarian lowlands and defended its territory against various attackers from the west.

In the 9th century A.D., the ancestors of the Czechs were united in the Great Moravian Empire. This was the first sovereign state structure on the territory of the modern Czech Republic. The territory of the Empire included today's Bohemia, Slovakia, southern Poland and western Hungary. The affairs of the state were managed at that time by ruling princes (Mojmir, Rostislav), selected representatives of the nobility and appointed servants. Influence of the church representatives in running the princedom was also considerable.

The culture of the Great Moravian Empire exerted a notable impact on the development of culture and religion among the Eastern and Southern Slavs in the Middle Ages. In 863, the Byzantine Christian missionaries Constantin and Methodius came to Moravia to introduce Slavic liturgy and Cyrillic alphabet. Later, however, the influence of the Roman Catholic Church gradually expanded and became dominant. The Great Moravian Empire collapsed following invasions of the Magyars in the years 903-907. It is noteworthy that a significant part of the local aristocracy remained more or less undisturbed by the fall of Great Moravia and their descendants became nobles in the newly formed Kingdom of Hungary.

The Premyslid and Luxembourg Dynasties

During the reign of the Premyslid dynasty (10th–13th centuries), Bohemia became the center of an independent state-building process, gradually grew in strength and succeeded in preserving its actual sovereignty despite formal vassal ties to the Holy Roman Empire. As a result of the strife for independence the foundations of local self-government were also laid down during this period.

In 1212, Prince Otakar I received The Golden Bull of Sicily, a decree proclaiming Bohemia a kingdom and Bohemian princes hereditary kings. Bohemia then became one of the most important states within the Holy Roman Empire. The reign of the Luxembourg dynasty began when John of Luxembourg (1310-1346) was elected King of Bohemia in 1310. The Luxembourg kings added new regions to their kingdom which received the name the Crownlands of Bohemia.

The kingdom of Bohemia reached its height of power and prestige during the reign of Charles IV (1346-1378), the second Luxembourg on the throne of Bohemia. In 1348 he established Charles University - it was the first university founded north of the Alps.

The Hussite Revolution

The late 14th and early 15th centuries were marked by the economic and political crisis during the reign of Wenceslas IV (1378-1419), the successor of

Charles IV. This period also witnessed an influential Church-reform movement, the Hussite Revolution, led by a Czech Jan Zizka. The latter was inspired by the teachings of Jan Hus, a preacher who was burnt at the stake in 1415 at Constance.

Heir to the crown of Bohemia, the Roman Emperor Sigismund, tried to crush the growing revolution by force but the Hussites defeated his five consecutive crusades in the years 1420-1431. Only the victory of the Emperor's troops in 1434 opened the way for a temporary agreement between Hussite Bohemia and Catholic Europe. The Hussite movement contributed to the changes in the structure of society and establishment of religious dualism for the first time in Christian Europe. The power and prosperity of the Church declined. The Czech nobility, towns and culture benefited from these developments.

The Habsburg Dynasty

In 1526 the Habsburgs of Austria succeeded to the throne of Bohemia. During his reign Rudolf II (1576-1611) left Vienna for Prague and the Bohemian capital grew into an important center of European culture. The Czech Estates forced Rudolf II to issue a decree proclaiming freedom of religious confession. The Emperors Matthias and Ferdinand tried to limit this freedom and their efforts sparked a civil war between the Estates and the Catholic Emperor. The Estates were defeated in 1620 at the Battle of the White Mountain. The following period of the Thirty Years' War brought political disorder and economic devastation to Bohemia with far-reaching implications for the future development of the country.

Bohemia lost its independence and became part of Austria and later of the Austro-Hungarian Empire for the subsequent 300 years. The people of Bohemia were forced to accept the Catholic faith or to emigrate. The throne of Bohemia was made hereditary in the Habsburg dynasty and the most important offices were transferred permanently to Vienna.

This period gave rise to alienation and resentment of citizens towards state officials who represented foreign power, spoke foreign language and acted often in an authoritative and arrogant manner.

A crisis of feudalism and the fiscal interests of the state led to the Enlightenment reforms of Maria Theresa and Joseph II in the second half of the 18th century. The reforms brought more autonomy to Bohemia and Moravia. At the same time the reforms contributed to the centralization of power and to the process of Germanization which proved to be a serious threat to the identity of the Slavic nationalities of the empire.

Bureaucracy played a key role in public life of the Habsburg Monarchy and especially in resolving various problems caused by multinational structure of the Empire. "Bureaucracy dominated the government under the Hubsburg Monarchy" (Liegle and Muller, 1999). Development of self-governed society was not encouraged since local self-government bodies were largely based on national affiliation and the central state suspected them of disloyalty. The whole system of education was oriented towards teaching students to be loyal citizens.

Formation of the Modern Czech Nation

Although the Czech national revival movement aspired at first only to a revival of the Czech language and culture, it soon began to strive for political self-determination. In the revolutionary year 1848, Czech politicians made the first coherent political proclamations aimed at transforming the empire into a federalist state. Plans for national revival were supported by quick industrialization of Bohemia which made the country the most developed monarchy in the second half of the 19th century.

The roots of many current administrative traditions, attitudes and behavioral patterns prevailing in the Czech society go back to the period when Bohemia was part of the Austro-Hungarian Empire. As one of the observers remarked "If Czechs have a penchant for bureaucracy (and by God, they do), then it is perhaps above all, a legacy of the remarkable social and economic modernization of Czech lands in the late 19th century". During this period "...they also developed a rich administrative vocabulary (most of it calqued from the German of their Austrian Hapsburg rulers) and a magnificent culture of bureaucracy and administration" (Hanley, 1999).

Independent State (from 1918)

In the years during World War I Czech politics took a turn towards radicalism largely as a result of the activities abroad of the future presidents T. Masaryk and E. Benes. The defeat of the Austro-Hungarian Empire cleared the way for the formation of an independent state of the Czechs and Slovaks (October 1918). The Constitution of the Czechoslovak Republic defined the new state as a parliamentary republic. In many ways it provided for a dominant position of the parliament over the Head of state and the Government. This was a reaction to the previous experience of authoritative regime. The political parties represented in the Parliament had excessive powers over the executive and to a considerable extent were determining the personnel policies of the state administration.

During the period between World Wars I and II Czechoslovakia had a developed parliamentary democracy. The state administration embraced almost unchanged the civil service system of the Austro-Hungarian Empire. The latter was structured largely in accordance with Weberian principles of developed bureaucracy and respective legal norms (Civil Service Rules of 1914 – Act N 15 of the Imperial Code of 25 November 1914). The state administrative structure comprised two institutional levels: 1) the organs of the state i.e. central government and other state authorities including those at the regional level; and 2) the organs of territorial self-government at the level of the lands, districts and localities.

During the inter-war period the country became one of the ten most developed countries of the world. This period of twenty years of democracy and prosperity was ended by the aggression of Hitler's Germany. The conference in Munich and the following German occupation in March 1939 brought an end to the independent Czech state.

The Period of Communist Rule

After World War II, the restored republic became part of the Soviet sphere of influence. A period of "limited" democracy was ended by a communist takeover in February 1948. Much private property was expropriated and political and human rights were suppressed. The civil service was formally abolished by the Labour Act of 1950 which did away with the distinction between public and private employment.

The communist regime subjected state administration to strict ideological control and subordinated it to the monopoly power of the Communist Party of Czechoslovakia. State administrative structures served as one of the main instruments for implementing political and economic decisions of the ruling party as well as interference in all fields of life both social and individual. Employment to any more or less significant post in state administration required demonstration of evidence of political loyalty, acceptance of the principles of "democratic centralism" and authorization of the respective communist party organ.

In practice, the so called principles of "democratic centralism" defined the mechanism of internal organization and management within state administration. In accordance with these principles democracy functioned within narrowly circumscribed limits set by party leaders. When the party elite made decisions, the rank and file members were obliged to endorse and strictly follow these decisions.

The communist regime downgraded the position of administrative staff. It abolished the legal status of civil servants which was effectively replaced by regulations of the Labour Code. The formal distinctions between an employee in a public sector (factory worker, teacher, doctor etc) and a civil servant ceased to exist. Administrative management was exercised by a system of the party nomenklatura whose representatives occupied positions at all levels of administration.

In 1968 the "Prague Spring" marked an attempt to "humanize" the country's political system, its state apparatus and above all its relations with citizens. In April 1968 Action Programme of the Communist Party declared "democratization" of the political and economic system as its main objective. The reform programme called for abolition of censorship, free travel abroad and also for replacement of incompetent officials and closer involvement of the public in decision-making process. The "Prague Spring" was suppressed by Warsaw Pact troops in August 1968.

Although plans to reform the political and economic system of Czechoslovakia in 1968 failed, many ideas and slogans popular at that time were reactivated after the victory of the "velvet revolution".

"Velvet Revolution"

The gradual decay of the Communist rule and the Soviet influence culminated in the overthrow of the Communist regime in November 1989. Following the "velvet revolution" the first free elections were held in 1990. Vaclav Havel was elected President of the Republic.

On 1 January 1993 the Czechoslovak state was peacefully divided and the independent Czech and Slovak Republics were founded. The Czechs initiated the process of the transformation of the formerly centralized state system into a parliamentary democracy and market economy. On 1 May 2004 the Czech Republic became a member of the European Union.

Summing up, we may say that the modern history of the Czech Republic has been a history of predominantly non-democratic political regimes with only one short exception – the period between the two world wars. The roots of the Czech state administrative system arise from the traditions of the Austro-Hungarian bureaucracy and principles of state management as practiced by the ruling communist party. It is interesting that, over the centuries, the prevailing in Czech lands bureaucratic attitudes and patterns of behavior became the source of great literature such as Franz Kafka's 'The Castle' and Jaroslav Hasek's 'Good Soldier Sveik'.

Most typical features of the previous administrative systems were wide spread patron-client relations and high extent of politicization of the state administration. It is noteworthy that even during the period of the Czech independent state between the two world wars "the political affiliation of an applicant for a civil service post played an important role in recruitment decisions" (Vidlakova, 2000:8).

5.3 Constitutional Framework

The Czech Republic is a parliamentary republic. It is defined by the constitution as a sovereign, unitary and democratic state governed by the rule of law based on respect for human rights and freedoms. The constitution states that the people are the source of all power in the state. They exercise it through respective bodies of legislative, executive and judicial branches of power.

The new Constitution of the Czech Republic (Constitutional Act of the Czech National Council No. 1/1993 of 16 December 1992) came into force on 1 January 1993. In accordance with another Constitutional Act of the Czech National Council No. 4/1993 of 15 December 1992 all laws and other legal regulations of the Czech and Slovak Federal Republic that were in effect on the day of the dissolution continued to be binding unless they were cancelled. The Charter of Fundamental Rights and Freedoms is an integral part of the Czech constitutional order.

The Constitution of the Czech Republic defines general authority of state administrative bodies and serves as the main legal basis for its activities. The extent of state executive power is defined explicitly in Article 2: "State power shall serve all citizens and may be applied only within the limits and in the manner provided for by law."

The legislative power belongs to the Parliament which consists of two chambers: the Chamber of Deputies with 400 deputies elected for four years period and the Senate with 81 senators elected for a term of six years. One third of the senators are replaced every two years. The Constitution established a Constitutional Court consisting of 15 judges who are appointed by the President of the Republic and approved by the Senate (Articles 83-93). Judges serve ten-year terms. The constitutional court also rules on the abolition of laws or their individual provisions that are in violation of a Constitutional Act or a ratified international treaty on human rights and fundamental freedoms.

The Czech President is formally elected by the Chamber of Deputies and the Senate for a maximum of two five-year terms. He is the Head of State but his role is largely ceremonial and his real powers are rather limited. He heads the armed forces, nominates the Prime Minister and in special cases can dissolve the lower house of the Parliament. He approves laws and has a power of veto though this can be overridden by a simple majority of deputies. The President cannot announce referenda or rule by decree. Because of the personality of V. Havel, the President in the Czech Republic has been widely perceived as a moral guardian of the nation.

In practice the balance of power between the President, Government and Parliament is gravitating towards the Government which defines the main directions of the country's internal and external policy. A survey conducted by Johansen among the Czech officials and experts revealed that in the Czech Republic the Government is considered to be the most important forum for decision making (Johansen, 2004:2).

The first constitutional change after the demise of the communist regime was made by the Constitutional Act of 3 December 1997, "On the Establishment of Higher Territorial Self-governing Units and on the Change of the Constitutional Act No.1/1993 - the Constitution of the Czech Republic". This legal act provides for the division of the Czech Republic's territory into communities which are the basic territorial self-governing units, and regions which are the higher territorial self-governing units. As a result 14 regional governments were established.

5.4 Political System

As we noted above effective political power in the Czech Republic is concentrated in the hands of the Prime Minister and his Cabinet who defines the main directions of governmental activities and priorities of drafted and reviewed legislation. As a collegiate body the government is responsible to the Parliament (Chamber of Deputies). In particular, it is accountable to the Chamber for the implementation of the state budget.

Elections to Parliament are held through a proportional representation system based upon party affiliation. The minimum level of popular support required to qualify for representation in Parliament is 5% of the local vote. The largest four political parties are Civil Democratic Party (ODS), Czech Social Democratic Party (ČSSD), Christian Democratic Union – Czech Folk Party (KDU-ČSL) and the Communist Party of the Czech and the Moravia (KSČM).

The development of state administration in any country is influenced by various factors – political, economic, historical, cultural etc. Three political factors deserve in our view special attention since they have had a considerable impact on the development of the Czech state administration throughout the 1990s:

- the process of formation of a democratic society and market oriented economy (started in 1989);
- the break-up of the former Czechoslovakia into two independent states (1 January 1993);
- the preparation of the country for accession to the European Community (since 1995).

In the opinion of some observers the Czech political system has been crippled (at least in the 1990s) by two essential drawbacks. Firstly, the major political parties have not enforced minimum accountability and quality by serving as watchdogs of each other's conduct and activities. Secondly, too often the system displayed a high level of tolerance towards politicians who failed to govern competently and honestly.

In his speech to both houses of parliament in December 1997 President Vaclav Havel emphasized: "The prevailing opinion is that it pays off in this country to lie and to steal; that many politicians and civil servants are corrupt; that political parties – though they all declare honest intentions in lofty words – are covertly manipulated by suspicious financial groupings" (CTK, 9 December 1997). In particular he pointed to lawlessness, corruption and utter neglect of any moral principles that accompanied mass scale privatization in the country. In his words the situation in the Czech Republic was such that to respect the law meant to put oneself at a disadvantage to the privileged whose wealth has been so often built on violation of laws and ethical principles (ibid).

5.5 Economic Situation

The economic development of the country in the 1990s is often divided into two phases. The first phase is confined to the period 1990-1996 during which the Czech Republic was regarded as a success story in the region with the lowest rates of inflation (9.1%) and unemployment (2.9% of the labour force). The high GDP growth rate was accompanied by a relatively low budget deficit. The priorities of the economic policy at that time included:

- price and trade liberalization;
- reduction of subsidies to enterprises;
- transfer to full currency convertibility;
- rapid privatization programme;
- restrictive monetary and fiscal policy.

During this period elementary institutional foundations of the market economy in areas such as property rights, the financial sector, corporate governance and bankruptcy were laid down.

The second phase 1997-2001 was characterized by decline in investment, real GDP growth at 0.3% and depreciation of the Czech Crown by 20% against USD. By the end of the decade the Czech economy was in deep recession. Unemployment doubled reaching 8.7% at the end of 1999, major banks tottered on the brink of collapse and industrial production was plummeting.

The major reasons for this unsatisfactory performance were slow pace of enterprise restructuring, growth of real wages above productivity, unconvincing results of voucher privatization, inadequate enforcement of legal norms and, in particular, protection of shareholders' rights.

Opinion polls provided further evidence of an "unhealthy" state of affairs – from 31 per cent among the Czechs in 1995 who thought the reform process had gone well the number of positively minded respondents had fallen to a mere 5% by 1998 (Shepherd, 2000:76).

An appropriate illustration of the situation was provided by one of the observers: "The culmination of all of these problems within the Czech economy created an atmosphere of failure in the government. In 1997, as the reform process appeared to be coming to a close, several drastic measures were introduced to keep the economy and the nation's infrastructure from falling well below its forecasts. Despite the efforts made by the government and private firms, the Czech economy still faltered heavily and to a great extent many of the causes of the financial crisis could have been prevented in the early phase of transformation. Vaclav Klaus' declaration that the transformation process in the Czech Republic had been completed was premature by any measure." (Kampars, 2004:41).

Radical reform measures taken by the Czechoslovak government in the initial years of transformation came at a very high social cost. According to official government statistics, in the first three years of transformation (1991-1993) the GDP in the Czech Republic declined by 21.8 percent; industrial output dropped even more by 34.8 percent. (Adam, 1999: 4).

Statistics published by the Czech Statistics Office in 2000 indicated that the Czech economy started to gradually emerge from its three-year recession. In this connection an OECD report argued that the Czech case shows a lesson "about sequencing of reform and the importance of establishing a sound regulatory and institutional framework to allow the market to operate". (OECD, 2001:160).

5.6 Privatization and its Implications

In the early 1990s the economic aspects of reform were considered by the Czech politicians to be of highest priority while the importance of introducing regulatory mechanisms was largely ignored. Development of institutional and regulatory framework was lagging far behind the process of privatization.

Numerous sources indicate that absence of a sound legal framework allowed the dishonest to misappropriate huge assets from the state. Weak control and easy access to huge financial resources contributed to wide proliferation of corruption and economic crime.

Selling off of formerly state-owned companies (coupon privatization) was launched shortly after "velvet revolution". Many state monopolies like those in the telecommunications, automobile and tobacco industries were privatized. This was not the case with regard to the majority of state banks. A debt structure retained by the formerly state companies from communist times led to the emergence of a complicated system of mutual insolvency. The state-run banks were unwilling to foreclose on virtually bankrupt companies fearing an increase in unemployment. At the same time banks tended to offer limited loans at very high interest rates. When mismanagement and embezzlement caused a number of leading Czech banks to fall in the mid 1990s, the government bailed them out instead of allowing them to go bankrupt. Later the government repeated the mistake by not letting any large enterprises to fail on account of their unmanageable debts.

The government support to medium and small businesses was not convincing though the development of the private sector needed effective assistance after 40 years of communist rule. Taxes for both individuals and business were rather high. In 1996, restrictions on trading of the Czech crown were lifted, making it the first convertible currency in the former Soviet Bloc.

In mid 1997 negative trends in economic development and foreign market speculation with the newly convertible Czech currency brought about a monetary crisis. All the deficiencies in the Czech privatization process and economic transformation were glaringly disclosed.

As one observer noted: "Economic privatization was expected to carry considerable costs, but dispersed equally across the population. Instead, not only are these costs perceived to be unequally distributed, but they are also considerably higher than first expected" (Hallstrom, 2005).

An opinion poll conducted in early 1998 revealed that over 50% of the Czech population felt they had been deceived by coupon privatization while more than two-thirds thought it had benefited only the government and the dishonest (Shepherd, 2000:104).

5.7 Legal Framework

The Constitution of the Czech Republic stipulates (Art.79) that "The legal relations of state employees in the ministries and other administrative offices shall be regulated by an Act of Parliament". This is the only provision of the Constitution directly concerning personnel employed by the Government and other state institutions. The Constitution also states that ministries and other administrative structures may be established and their competences defined exclusively by an Act of Parliament (ibid).

According to the Czech legislation a state institution has to be established by a legal act. The Law on Competences (No. 2/1969) comprises a list of ministries and determines their functions and authority. This law defines the overall division of tasks within the central administration and its basic organisation. Administrative procedures are governed by the Law on Administrative Procedure (No. 71/1967) and some other laws.

In the Czech Republic neither the Constitution nor the Law on Competencies established clear division of authority between ministries and other governmental institutions. There has been a shortage of conceptual and legal definitions of organisational entities such as Ministries, Committees, Agencies, Inspections, etc. This vacuum resulted in sometimes confused perceptions of what should be part of the political mandate of a Minister and what should not. (Fernandez, 2001).

Until 2002 the civil service in the Czech Republic did not exist from the legal point of view since state administration employees (as one whole) were not covered by special regulations. Unified standard criteria for civil servants selection, recruitment, promotion and management were missing. As a rule each government agency defined its own policy in this respect. A special law on civil servants was drafted by the Ministry of Labour and Social Affairs and submitted to the government in September 2000. It was approved by the Czech Parliament in April 2002 (Civil service Act No. 218/2002).

Approval by the Parliament on 21 March 2000 of the Act 150/2000 on Judicial Administrative Proceedings (enforced from 1 January 2002) created a system of administrative justice in the Czech Republic. There are two areas of jurisdiction of the Administrative court: appeal to specialized chambers in regional courts and cassation before the Supreme Administrative Court (set up in Brno in 2003). The courts have the right to replace or annul administrative decisions. However, due to limited experience it is too early to speak about their effectiveness.

5.8 Central Government

The Government is the central policy-making and executive body. It consists of the Prime Minister, Deputy Prime Ministers and Ministers. The number of Deputy Prime Ministers is not determined by law. In 2006 the government consisted of 18 members, including the Prime Minister, two Vice Prime Ministers (one of whom also acts as Minister of Labour and Social Affairs), 14 ministers in charge of separate ministries, and a minister without portfolio. The ministers are appointed and dismissed by the President on the proposal of the Prime Minister. The majority of ministries have regional offices.

The Prime Minister chairs Government meetings which are held on a weekly basis. Decisions may be passed by a majority of members present. The Government issues decrees within the scope of existing law. Decrees must be signed by the Prime Minister and the government member concerned. Ministries, other administrative agencies and territorial self-government bodies may issue legal regulations on the basis of existing legislation and within the area of their authority.

The Office of the Government is an advisory and administrative support structure providing information, technical and organizational assistance for the activities of the government. It is a central institution which does not fulfill ministerial functions.

The line ministries are responsible for designing and implementing government policy in a specific area. Their internal organisation is decided by each minister. Formal guidelines and mechanisms for policy coordination among ministries are generally in place and are administered by a special department in the Cabinet. Policy coordination is usually achieved through government advisory structures created by a decision of the Government and by ad hoc commissions formed by separate ministries.

One of the examples of the latter was the Inter-ministerial Commission on Public Administration Reform established by the Ministry of Interior. The Commission consisted of Deputy Ministers of all ministries and Heads of selected central administration bodies. It was headed by the Deputy Minister of Interior in charge of public administration reform.

Various reports show that although mechanisms for inter-ministerial co-ordination are in place they are generally weak and the procedures are slow. Negotiations on draft laws, on policy-proposals and on budget lines take a long time and are not settled at appropriate levels. Also ministries, departments and sub-departments often have over-lapping functions and communication and management structures are blurred. Effective elaboration and coordination of policy at the central level is complicated by the tendency of line ministries to focus on narrow sectored interests at the expense of broader cross-sector priorities. As one expert stated "Each ministry is an isolated fiefdom working independently with little cooperation or sharing of information" (Chandran. 2000:10).

5.9 Civil Service: Status, Efficiency and Institutional Setting

In the 1990s the Czech Republic has undertaken certain reforms aimed at developing institutional environment to support competitive markets and democratic values. Progress has been made in devolving powers to local and new regional governments, introduction of certain regulatory mechanisms, and harmonization of Czech laws with EU legislation.

Little attention however was paid to the modernization of the state administration. The need for reform of public administration was regularly pronounced. However, the reform 1) was conceived almost exclusively as reform of regional and local government; 2) its implementation was permanently delayed; and 3) even when modernization of central state administration was considered, it was perceived mainly as adoption of respective legislation. "The public administration reform has been understood in the Czech Republic in a very narrow way as the setting-up of regional authorities and the adoption of the Civil service Act" (NTF, 1998: 7-1-7).

A particular feature which distinguished the Czech Republic from many of its CEE neighbours was absence until 2002 of legislation providing for the existence of a civil service and respectively lack of formal status of a "civil servant" as such. Each employee had a contractual relationship with a specific institution (ministry, agency etc) but not with the state as a whole. The Labour Code served as the basis for this relationship.

The problem remained, however, that whereas by the early 2000s the number of civil servants had increased, the quality of civil service had fallen. Speaking to the deputies of the Parliament, the Czech Deputy Prime Minister remarked "The level of civil service in the Czech Republic since 1989 has dropped considerably" (Spidla, 2001).

The functions of central government bodies were not sufficiently clarified within the general system of public administration. The role of civil servants in the policy process was not clearly defined either. The internal regulations of ministries governing their structure and responsibilities of their units differed considerably in their content and extent. Typical of the Czech institutional setting has been weak law enforcement. "As with all laws in the Czech Republic, enforcement is a concern" (Chandran, 2000:11).

Promotion and appointment based exclusively on merit were rather an exception than a rule. Unethical and corrupt behaviour of civil servants continued to arouse serious concern. "The administration is still politicized and lacks sufficiently qualified staff to serve as reform agents" (SIGMA, 2003).

In spite of these negative trends the Czech Government "did not initiate any in-depth analysis of this state of affairs and did not deal systematically and conceptually with the problems of civil service in general or the personnel policy of state administration in particular" (Vidlakova, 2000:8).

An OECD report described the situation in the following way "Today many of the highest priority challenges facing the Czech Republic lie in improving the capacities of the public administration – its skills, its structures, its accountability for performance, its relation with and understanding of markets and consumer interests, its culture and its style of operation" (OECD, 2001:136).

The period of the 1990s revealed also that in the Czech Republic there was no influential political party or social group, which not only displayed interest in civil service reform but was committed and motivated to push for its implementation.

5.10 Civil Service Management

Since the institution of civil service was established in the Czech Republic only recently and the basic law which determines largely its activities has not been fully enforced yet, it is possible to speak here only about the way state administration was managed before and the way it should be managed according to the new legislation.

As we pointed out already, prior to 2002 management of the Czech state administration was largely compartmentalized and exercised mainly by the leadership of each ministry or agency. Universal standards of the organisation, recruitment, assessment, motivation of civil servants were lacking. There existed general provisions in the Constitution that overall management of the central administration is exercised by the government of the day. The terms of service were regulated mainly by the Labour Code.

Management of state administration was characterized by insufficient coordination, orientation on process rather than results, disregard of the importance of proper training and considerable neglect of citizens as ultimate clients.

The level of decision-making was excessively "top heavy" with 90% of the decision-making powers concentrated on average in around 6% of employees of the state institution (NTF, 1998: 3-1).

The new law adopted in 2002 envisaged that civil service management would be exercised by the General Directorate for Civil Service which was established in June 2002 as a separate unit in the Office of the Government (SIGMA, 2003). The law also provided for the introduction of a new post in each ministry and agency - Director of Personnel who is reporting directly to the General Directorate for Civil Service.

However, full enforcement of the Civil service Act has been regularly postponed. Four years after the adoption of the Law the post of the Head of the General Directorate for Civil Service was still vacant and the Directorate's staff was cut by 60% (Landova, 2006:60).

5.11 Relations Politicians - Civil Servants

Until the adoption of the Civil Service Act a clearly defined distinction between political appointees and tenure government officials was absent in the Czech Republic. The legal framework guiding relations between politicians and civil servants has been fragmentary. Some elements of this framework were incorporated in the Conflict of Interest Act (No. 238/1992) and partly in the amended Labour Code (No. 65/1965).

After the fall of communism in 1989 politicization of the state administration in the Czech Republic continued to be a problem. An analysis prepared by the National Training Foundation stated "We have also failed to get rid of the politicization of administrative staff.... Some officials believe that the most important thing is to be loyal to a certain political party, which can bring more profit than perfect performance of the service. Czech politicians have not grasped yet how such an approach can damage public administration and, consequently, also themselves through undue influence upon state officials". (NTF, 1998: 6-2).

One of the objectives of the Civil service Act adopted in 2002 consisted in reducing politicization of the civil service personnel. However, since the full enforcement of the Act is still pending, the relationship between politicians and civil servants in the Czech Republic continues to be based on many stereotypes of administrative culture and conduct formed during the communist and pre-communist past.

The negative impact of excessive politicization of the administrative apparatus was noted by many observers. "The first weakness in the Czech bureaucratic structure is that there is no clear separation between civil service and politics. Civil servants come and go with changing governments. Thus, the main agenda of the civil service is to serve the ruling party rather than the people" (Chandran, 2000:10).

"While personnel changes within ministries due to changes in the cabinet could hardly be said to represent anything unique in the world, it needs to be emphasized that in the Czech Republic these changes often affected even the lowest management levels within the institution; something that would be impossible in many democratic countries due to the safeguards provided by civil service laws" (Landová, 2006:15).

Deputy Prime Minister Spidla introducing a draft Civil service Act to the Czech Parliament referred to "unhealthy politicization of the civil service". "It is becoming a rule that every change of government or a minister is accompanied by replacement of key staff (a sort of personnel earthquake) which has a very negative impact on the quality and effectiveness of work." (Spidla, 2002).

One can find even more critical generalizations "Current unsatisfactory status of the Czech state administration serves evidence of the negative effects of the long term political interference in the management of public administration" (Onisko, 2000:36).

5.12 Human Resources Management

The main feature of personnel management in the Czech state administration has been its decentralization. Data on the number of employees in the Czech public administration coming from various sources is incomplete and conflicting. Since the formation of the Czech Republic in 1993 the number of state administration employees has been growing. In 1998 there were 13,000 civil servants in the central state administration. In 2001 the number has reached almost 17,000. Overall number of state administration employees in 1998 amounted to 154,000 (World Bank, 2000:122).

According to Williams the Czech bureaucracy doubled in size under the "pseudo-Thatcherite" government of Vaclav Klaus (Williams, 1999).

Prior to 2002, there was neither a centralized human resources management system dealing with employees in the state administration, nor standard criteria for civil servants' selection, recruitment, promotion and management. The establishment of such a system was envisaged in the Civil service Act full enforcement of which has been pending for years.

In fact, in matters of employment and personnel management, every institution has acted as an independent legal entity and not on behalf of the State or on the basis of State authorization. Even more important, strict observance of merit based principles in appointments, promotion and compensation was rather an exception than a rule. "Practices in recruitment and promotion are not based on open, merit based competition, with the result that they encourage political allegiance and patronage and put the professionalism and impartiality of the civil service at risk" (SIGMA, 2003:15)

The workload of individual civil servants has been generally unbalanced. Those who are capable are overburdened and the administration finds it difficult to dispose of those who are incompetent and inefficient under the existing conditions (NTF, 1998: 6-2).

Insufficient prestige of civil service and relatively low levels of remuneration weaken the ability of the administration to retain employees. As a result, staff turnover has been high, particularly of young specialists who have been employed by the government only for a year or so. Due to the same reason existing training facilities have been overburdened.

There has been no effective policy of making public sector career attractive for qualified and experienced entrants. Public and professional recognition of service to the government and attractive career prospects are largely absent. "Civil servants in the Czech Republic have neither job security nor a clearly defined career path which discourages young people from entering the civil service" (Chandran, 2000:10).

Mobility between ministries or from outside the public service has been limited which reduced exchange of experience and new ideas. "There is a shortage of professionals with general orientation, modern education for management positions and strategic decision-making as well as of members of a number of professions and specializations, trained for performing analytical, information, legislative etc. functions in public administration." (NTF, 1998: 6-2).

5.13 Pay and Incentive Mechanisms

Until the adoption of the Civil service Act civil servants' remuneration system was based on the Act on salaries and remuneration in budgetary organizations (No. 143/1992) from the year 1992. This act defined salary scales, terms for granting bonuses and extra pay and determined procedures for management of payroll.

Principal elements of this system (NTF, 1998: 5-5) comprised:

- the core of the salary model is formed by the base pay (tariff);
- salaries are based on salary classes and grades according to the number of years in service;
- there are personal evaluation bonuses which should reflect the performance and achieved results over a longer period of time;
- additional allowances;
- extra pay reflecting specific conditions for the performance of work and the nature of work activities.

The bonuses for the grades 9 to 12 and 1 to 8 could amount up to 100 and 40 per cent of the base salary respectively (SIGMA, 2002:7).

Modern staff appraisal and evaluation systems, proper job descriptions and clear rules for career advancement have been largely absent thus allowing strong discretionary powers of superiors. The mechanism of supplementary pay (bonuses and other discretionary allowances) lacked clarity and transparency and resulted in significant differences in take-home pay between officials of state institutions. Such practices created a demotivating environment restricting efficiency, quality and initiative of rank and file civil servants. (NTF, 1998: 5-5)

In the late 1990s the Czech Ministry of Finance introduced a policy of providing an overall wage budget to line ministries while allowing substantial discretion in its use, subject to compliance with existing regulations regarding remuneration (World Bank, 1999:4). Such practice was aimed at ensuring effective management of human resources at the same time preserving overall fiscal discipline. However, absence of effective mechanisms for holding budgetary units accountable for how effectively the line ministries attain policy and plan objectives undermined motivation to effectively exercise this autonomy.

The comparison of salaries in comparable jobs between state administration and private sector shows that the average income in state administration for mid level staff is lower by 30% or more (NTF, 1998: 5-5). Until late 1996, the earnings in the public sector, in particular of employees performing more highly qualified, complex and responsible work, were generally on the rise. However, in 1997 the government austerity measures resulted in the drop of remuneration in the state administration in contrast with the business sector. These developments led to high turnover of personnel, a tendency to seek higher income elsewhere and proliferation of corruption.

The Civil service Act (if fully enforced) should provide for an overall higher level of remuneration of civil servants and closer dependence of take home pay on the results of performance appraisal. Still, OECD experts assessed the proposed new salary and supplementary pay system as "rather complex and fragmented, which may give rise to some interpretational and operational difficulties". They also did not exclude difficulties in budgetary funding of this expenditure (SIGMA, 2002).

5.14 Education and Training

Throughout the 1990s neither specialized university level education nor systematic in-service training of central state administration officials existed in the Czech Republic (Vidlakova, 1999:11, Potucek, 2004:97). Upgrading qualifications and re-training of civil servants had been managed *ad hoc* by many central and regional public administration institutions without any coordination. The role of training in the resolution of urgent problems of public administration development was reduced to mostly random sending of individuals to training courses (including courses and study abroad). No central authority for coordination of the training of public administration employees existed at that time.

A central higher education institution specializing in a targeted preparation of civil servants and their training did not exist either. Motivation for training and self-training has been overall weak. The level of education of 17% of civil servants in central state administration did not correspond to the qualification requirements of the post they occupied (Concept of Training, 1999). It can be generally stated that an opportunity to extensively utilize training as one of the basic tools of the reform and modernization of the Czech public administration has been missed. (NTF, 1998)

In 1996 the opinion of all ministries was requested regarding the necessity of establishing a Public Administration School of university level. The surprising result was that the majority of ministries was satisfied with the existing situation and did not consider the establishment of such a school a priority.

In June 1999 the Czech Government approved a policy for training civil servants. The main objective of this document was to establish life-long training in an open system providing qualified preparation of civil servants to standards appropriate to EU accession. The system would be provided by state and private sector middle-level and higher schools and universities. The document highlighted the necessity of establishing a central authority for personnel management to coordinate registration and professional education of state administration employees.

The Institute of Public Administration was finally established in June 2001 as a structure affiliated to the Office of the Government. Its current main functions are the provision of in-service training and coordinating training activities for the central government agencies.

5.15 Accountability and Responsibility

During communist rule for understandable reasons the notion of accountability or responsibility of state administration employees was regarded almost exclusively as accountability/responsibility to the ruling party or, in other words, to the leadership of the ruling party.

After the "velvet revolution" the Czech political leaders enjoyed legitimacy and support of the population. They prioritized purely political and economic aspects of the reform. The development of comprehensive regulatory frameworks and mechanisms of oversight and accountability was delayed.

It should be noted that there does not exist direct and precise translation of the concept 'accountability' into the Czech language (Konopasek, 2002:1). "Accountability" is usually translated as "odpovednost", but this word can be best equated to the English term "responsibility". A group of Czech experts who studied this issue proposed another expression "verejna vykazatelnost" which is in their view equivalent and more precise (ibid).

Difficulties with translation and historical heritage did not facilitate development of the appropriate environment and introduction of effective mechanisms of ensuring politicians' and civil servants' accountability above all to citizens and civil society. The situation was further compounded by overall immaturity of the Czech civil society and the prevailing conviction among state administrators that citizens exist for them, not the other way round. As a result "there has been insufficient public and civil control over the performance of public administration" (Marek, 1999), "the main accountability mechanism continues to be hierarchical subordination" (SIGMA, 2003).

In summary, we refer to the statement of one of the Czech experts in the field who asserted that the process of introducing real accountability in the Czech public administration "will be most difficult and most protracted" (Vidlakova, 2000:21).

5.16 Ethical Standards and Corruption

The Labour Code which was approved long before the "velvet revolution" but remained binding on state administration employees for more than decade after, contained certain restrictions concerning ethical norms and potential "conflict of interest" situations. These comprised requirements amongst others to ensure confidentiality of information obtained on the job, to maintain political neutrality, to refuse accepting gifts or favours in connection with one's post, to ask for permission in order to engage in private business activities (World Bank, 1999:8). Even so, these quite appropriate measures were not supported by sufficient institutional arrangements to ensure compliance.

In general, issues of ethics received relatively little attention in the Czech public administration (NTF, 1998 p.6-4). There was no separate code of conduct for civil servants in the Czech Republic until 2001. Code of Ethics came into effect in 2001 but "it is vague and largely repeats provisions already stated elsewhere" (Open Society Institute, 2002).

The Conflict of Interest Act (N 238/1992) was adopted in 1992 but its applicability was confined only to the ministers and heads of other central state administration institutions. The Act in particular required that these top level officials should regularly submit reports on their (and their spouses) personal income, real property and other assets (Vidlakova, 1999:6).

As we have underlined above proliferation of unethical behavior and fraud was facilitated by the relatively low level of remuneration of Czech civil servants, absence of a sound regulatory framework and wide opportunities for graft and embezzlement that opened up in the course of privatization.

Attempts to reduce red tape and corruption have often failed because of inconsistent political backing and resistance of bureaucracy. For example, the Prime Minister's Office advisory body on removing bureaucratic burdens was dissolved in summer 1998 after only one year of operation (World Bank, 1999:98).

The considerable degree of corruption in the Czech state administration is acknowledged by many observers. A report of the Czech Ministry of Interior on corruption submitted to the Government in 2001 stated that the number of bribes was not decreasing and not only civil servants but even representatives of political parties were engaged in corruption ("Mlada Fronta Dnes", 2001). Another report noted "the areas that appear to be seriously affected by corruption are the state administration, the legislative process, judicial system and public procurement. Political party finance appears to have receded as a corruption hot spot since the scandals of the late 1990's. Although the Czech Republic is not ranked as a country seriously affected by "State capture," corruption of the legislative process appears to be an increasingly serious problem, encouraged by uncontrolled lobbying, MPs' immunity and inadequate conflict of interest regulations. The dynamics of corruption have been shaped in very important ways by the nature of Czech privatization and its consequences" (Open Society Institute, 2002).

The CERGE-EI think tank identifies two dominant levels of corruption in the Czech Republic: a) high level corruption involving elected and senior public officials; and b) low level corruption involving mid and low level bureaucrats (Lizal and Kocenda, 2003).

Corrupt behavior has taken various forms. "Corruption is believed to be prevalent within the Czech civil service. Bribes are expected everywhere where some material benefit ensues, for example, in government procurement. Businessmen claim that no government contract is rewarded without a bribe of around 10 percent of the value of the contract. Bribes are also reported in applications for bank loans, registering transfers of property, receiving inheritance, restitution, privatization, registration of a business with a court, and university admission" (Open Society Institute, 2002).

The issue of anti-corruption strategy has attained particular significance after the government was forced to resign following allegations of the corruption in the V.Klaus' party - ODS. Czech citizens seemed to be frustrated by the fact that many scandals involving state and government officials had been so far seldom properly investigated and few officials had ever been punished.

In February 1999 the Ministry of Interior submitted an anti-corruption strategy under the title "The Government Programme of Fight against Corruption" which was approved the same year by the Government's resolution No.125. In spite of these efforts the results of the fight against fraud and corruption have been modest so far. In April 2002 the Czech Government approved its own report on corruption which confirmed that the situation was not improving. Overall, implementation of anti-corruption strategy remained a declared priority of subsequent reforms of the Czech state administration.

5.17 Societal and Administrative Traditions and Values

With regard to societal traditions and norms prevailing in Czech state administration one should bear in mind that in 1620 the Czech state lost its independence for over 300 years. Since that time the perception of state bureaucrats was very much associated with foreign officials and their local assistants appointed to run administrative matters. Generally this perception did not include local self-government which was considered by many Czechs as more loyal to citizens than central state administration.

Overall the Czech society is distinguished by the prevalence of egalitarian tendencies, weak religious feelings and general indifference to ultra-nationalism (Jalovecka, 2004:4). Ordinary Czechs have been often associated with such well known character as Hasek's Good Soldier Sveik and his high degree of conformism. Still widespread among citizens is the belief that government and administration should solve all problems including loss of jobs and closure of businesses.

Besides, contemporary Czech society is to a very large extent clientelistic and corporate. Groups of friends, acquaintances and relatives dominate most of the public sector (Tychtl, 1999:11). As a rule such an environment is favorable for preservation of patronage networks which are wide spread within the Czech political elite and civil service.

As we already know many traditions and behavioral patterns of the Czech bureaucracy developed under strong influence of the formal and informal norms and rules that prevailed in the state administration of the Austro-Hungarian Empire and later of the communist regime. Therefore these traditions are, on the one side, predominantly state centered and vested interests oriented, and on the other side - citizen unfriendly. Due to this historical legacy one often comes across the deeply entrenched custom of maintaining governmental and administrative secrecy in the Czech Republic, wide-spread cases of red tape, top heavy executive control and supervision and treatment of citizens as subordinates rather than clients.

Preparation for accession to EU and actual membership since 2004 have provided the Czechs with a lot of opportunities to become acquainted with best administrative practice in EU member countries and, what is more important, to improve compliance with EU average standards in this field. As a result for example cases of greater transparency and client friendly attitude have become relatively more common.

5.18 Civil Service and Civil Society

The relation of the Czech citizens to state institutions is characterized by a certain crisis of trust. The social distance between citizens and the state administration is still considerable and in some cases even critical.

There exist a number of factors that have led to such a state of affairs. On the one hand, the bureaucracy has largely discredited itself by numerous cases of corruption, frequent neglect of citizens' interests and overall attitude towards citizens as "inferiors". On the other hand, the Czech civil society is not strong enough yet to effectively defend its interests and there still persists widespread among the population paternalistic conviction that the state should resolve almost all the problems of ordinary citizens. The whole situation was compounded by obvious difficulties in the country's economic and social development, particularly in the second half of the 90s, which indirectly testified to the "efficiency" of the state administration.

The second part of the decade (after 1996) witnessed a sense of deep disappointment in society. Economic recession in 1997 complemented by corruption and political opportunism appears to have undermined the ideals of the "velvet revolution". "Klaus economic miracle had been exposed as a sham" (Shepherd, 2000:6).

When it turned out that Klaus's reforms were in many respects a failure, there was a radical change in people's attitudes: until mid-1996, more than 40 per cent of Czech voters were "very satisfied" with political and economic developments in the Czech Republic, and more than 30 per cent were "satisfied". In July 1997 it was only 12 and 8 per cent respectively. (Čulík, 2000).

Czech civil servants have often failed to establish good contacts with citizens, either due to a low degree of its professionalism and motivation or to their attitude towards citizens. In a survey conducted by the IVVM (research institute in Prague) in March 1998, 48% of interviewees stated that they had met with an unjust decision of the authorities. Negative experience was expressed primarily by people of productive age, with higher education, and by entrepreneurs. The majority did not have much hope in success in dealing with the authorities and enforcing their justified demands and rights. According to the survey, there often prevails a mistrust of citizens towards the government and ministries, and in some cases its extent is enormous. (NTF 1998: 4).

The lack of trust towards the administration is also caused by inconsistent prosecution of public offences, especially cases where high ranking officials are involved, and by the still rather ineffective mechanism of defending citizen's rights in case some damage has been inflicted by the authorities.

The early 2000s have seen some improvements in government policies towards the citizens. The Act on Free Access to Information was approved by the Czech Parliament and entered into force on 1 January 2000. The same year the Institution of Ombudsman was created in the Czech Republic. During the period 2000–2004 the Ombudsman dealt with over 14,000 complaints from the public and helped to resolve over 7,000 cases, providing legal advice to the remaining 7,000. Opinion polls conducted in 2004 showed that the Czech Ombudsman Motejl was the second most trusted public figure in the country. With a 53% approval rating he was second only to the President and considered more trustworthy than any cabinet minister including the Prime Minister (Radio Prague, 14 October 2004). Also recent years have seen much wider practice of placing information about the ministries' activities on their web pages.

5.19 Capacity and Resources

The need for institutionalization of the civil service and its modernization was acknowledged by all post-communist Czech governments. However, these declarations were not followed by consistent practical measures. There was obvious shortage of political will to adopt respective legislation and launch civil service reform. A document produced by experts working for a Phare project in the Czech Republic contained the following comment: "Reform of the public administration was accentuated in all the government's statements but has not been implemented so far."

Even when the start of public administration reform was actually announced it turned out that it had been substituted by the reform of the territorial government. "In the Czech Republic the professionalisation of the civil service and the construction of significant local autonomy have been self-consciously tied together across the party spectrum as issues of efficiency reform. However after ten years, this near cross-party insistence on connecting these two issues, combined with the persistent unwillingness to enact either, has looked more like a strategy for postponing losses of central party political autonomy than a coherent framework for developing efficiency" (Innes, 2002:32).

The post-communist period was also characterized by certain deficit of financial and human resources to prepare and implement reform of the state administration. The draft Civil Service Act is said to have been prepared by two part time experts who were occasionally assisted by the officials of the Czech Ministry of Labour and Social Affairs. The growing state budget deficit in course of recent years did not facilitate preparation and initiation of reform either.

One of the main factors that appear to have affected effective functioning of the Czech state administration has been inconsistent application of the meritocratic principles in human resource management. Performance based criteria have been either absent or had little value in the work of various administrative structures because of excessive politicization and prevalence of patronage networks. Even greater damage to the capacity and prestige of the administration seems to have been inflicted by the far from isolated cases of graft and embezzlement.

A condensed description of the capacity of the Czech state administration for reform and the conditions for its implementation was given by one Czech expert: "Public administration reform is prepared in the environment which is far from being favourable. Public opinion, politicians and, on the other side, public finances are not supportive of realization of the new law" (Onisko, 2000:5).

5.20 Conclusions

The development of the Czech state administration in the 20th and the beginning of the 21st centuries has been seriously influenced by the bureaucratic traditions of the Austro-Hungarian Empire and principles of administrative control and management as practiced more recently by the ruling communist party. It is precisely in this historical legacy that we can find the roots of such widespread phenomena as patronage and clientelistic networks, excessive politicization and secrecy, treatment of citizens as subordinates and frequent disregard of the interests of common people.

It is worth remembering that "The Communist regime in Czechoslovakia was amongst the most strict and conservative, particularly since it suppressed the "68 liberalization movement" (Toonen 1993: 159). In that sense the regime was very similar to its Soviet counterpart.

By the mid 1990s there had formed in the Czech Republic a model of central state administration with strong elements of a career civil service comprising lifelong employment, guaranteed promotion plan, complicated procedure of dismissal from the job, etc. However, this was a "specific" model since the status of state administration officials wasn't different from that of other public sector employees and the terms of service were governed by the Labour Code. There was no civil service related legislation and attempts to develop and adopt Civil service Act as well as to launch reform of the Czech civil service lacked sufficient political backing. "...little political priority was given to administrative reform" (World Bank, 1999:5)

Overall the post-communist period in the Czech Republic was marked by the absence of influential political party or social groupings that could effectively push for the start of civil service reform. Constraints in the development of a modern, efficient and accountable civil service were unfortunate as "the failure to recognize early the role of a reliable administration as an important element of an enabling framework of economic and political/social development continues to resound in the complex problems now facing the Czech Republic" (World Bank, 1999:5)

The main problems and deficiencies accumulated by the early 2000s in the Czech state administration were:

- persistent absence of civil service related legislation and hence absence of a legally binding status of a civil servant;
- excessive politicization of the state administration;
- domination of patronage networks;
- relatively high level of corruption;
- budgetary constrains in funding state administration;
- shortage of expertise in modern HR management;
- replacement of meritocratic principles by personal or party loyalty;
- lack of political will to "legalize" and reform civil service;
- underdeveloped performance based outlook and accountability to various stakeholder groups.

These "bottlenecks" of the Czech state administration were seriously weakening its capacity for effective functioning and reform. By the late 1990s it also became clear that the Czech state administration was substantially failing to meet the requirements imposed by the European Commission on the candidates for accession to the European Union. This was, in particular, reflected in regular evaluations by the European Commission on the progress made by the Czech Republic in the preparation for accession.

To conclude, it should be stated that throughout the 1990s the reform capacity of the Czech civil service was low across its all major parameters: institutional framework (absence of civil service de-jure, prevalence of outdated formal and informal norms and rules, preservation of many behavioral patterns, attitudes and traditions rooted in the communist and pre-communist past); collective or individual human factor (lack of political will of the country's leadership to launch reform, absence of influential stakeholders which could push for the "revival" of the Czech civil service); available resources (severe budget constrains and shortage of specialists in modern civil service management and reform).

CHAPTER VI
REFORM OF THE CZECH CIVIL SERVICE: INTERNAL IMPERATIVE OR EXTERNAL PRESSURE?

6.1 Introduction

In line with overall methodological approach we intend in this chapter to study various aspects and stages of the preparation and implementation of civil service reform in the Czech Republic.

For that we will need to trace the development of the Czech state administration after the "velvet revolution" (1989) and formation of independent Czech Republic (1993). We will analyze the history of attempts to launch reform of public administration and central state administration; identify reform dimensions, its dynamics and driving forces; examine its objectives as well as the process of reform implementation and management; and assess the preliminary results of reform if any. Particular attention will be paid to a set of issues relevant to civil service reform as defined in the Protocol for comparative studies of national civil service systems namely:

- who or what initiates reform and why?
- what are the contents of reform and up to date results?
- what is the extent of external influence?

Any researcher who undertakes to study recent evolution and changes in the Czech state administration should distinguish between *public administration (or administrative) reform* and *civil service reform (called reform of central state administration in Prague)*. The meanings of these terms have never been formally specified in such a way as to ensure unified acceptance and identical understanding by the majority of experts, scholars and civil servants. It should be noted that throughout the 1990s public administration reform was regarded by the majority of Czech government officials and experts entirely as reform of territorial self government and regional administration. "During the 1990s the public administration reform both in its conceptual and actual form concerned solely territorial self-government and territorial state administration" (Vidlakova, 2006:2).

Therefore, it is possible to speak about implementation of civil service reform in the Czech Republic in a more commonly perceived meaning only subsequent to 2004 when the Czech government approved basic documents concerning reform of the central state administration.

6.2 Preparation of Civil Service Reform

In spite of numerous declarations the reform of state central administration and central government remained outside the priorities of Czech politicians and legislators for many years after "velvet revolution" (Vidlakova 1999, Onisko 2000, Potucek 2001). It is true that the structure of the Czech government and central state administration was modified several times during this period (Vidlakova, 2006:2) but these alterations did not constitute part of a comprehensive reform measures.

Throughout the 1990s limited and sporadic changes that were taking place in the Czech state administration followed neither a properly developed strategy of civil service reform, nor were effectively managed by a specially designated management team or unit. An introductory note to the Czech government policy document in 2004 underlined that the Czech central state administration "has not been subject to any significant change since 1993" (Further advance, 2004:1).

Furthermore, throughout the 1990s the perception of civil service reform itself in the Czech Republic was confined almost exclusively to the preparation and adoption of the Civil service Act. Several attempts to draft a Civil service Act and get it approved were made over the period 1993-1996.

The first coalition government in the Czech Republic was formed in 1992. The coalition was led by Civil Democratic Party (ODS) and its leader V.Klaus. In its programme declaration the government defined the objective of eliminating old methods and rules in administrative practice and enhancing the professionalism of civil servants as well as efficiency and prestige of their work. Promise was made to prepare expeditiously a draft act on the legal status of state administration employees (civil servants) "defining the requirements imposed on these employees and their duties as well as certain compensating measures assuring stability and independence of their status" (Vonkova, 1996:120).

An important event took place in October 1992 when the Office for Legislation and Public Administration of the Czech Republic (OLPA) was established under the chairmanship of the Vice Prime Minister. From the very beginning this institution was assigned a significant role in drafting civil service related legislation and developing a conceptual vision of the future civil service. Regretfully, this office was disbanded in 1996.

In June 1993 the Minister of Labour and Social Affairs presented to the Government a draft document outlining the principles of Civil service Act which the Government approved in August 1993. The principles were based on the assumption that a career (tenure) civil service would be established in the Czech Republic. Principal drafting responsibility at that time rested with a small team in the Ministry of Labour and Social Affairs and the Office for Legislation and Public Administration (OLPA). Later this document was positively appraised by the respective Committees of the Chamber of Deputies of the Parliament. On the basis of the approved principles the full text of the draft law was elaborated and submitted to the Prime-Minister in July 1994. The draft comprised over 300 articles and "granted important privileges to civil servants" (Onisko, 2000:2). However the process of approval and subsequent revision of the draft became extremely difficult and time consuming. This draft was criticized in particular by the Ministry of Finance which referred to budgetary restrictions. In the end the draft was abandoned in 1995.

The second coalition government, headed also by V.Klaus, promised again to prepare a draft Civil service Act regulating the status of civil servants and ensuring its independence from political and party interference. A working group established within the Ministry of Labour and Social Affairs started its work but as a result of budgetary constraints in 1997 the activities of this group were suspended.

In March 1998, the newly formed government appointed to bridge over the period prior to extraordinary elections in June 1998, committed itself to public administration reform including measures aimed at the re-organisation of the central government. The main priority of the reform programme was the creation of the regional tier of government. In view of the short period of its existence this government failed to get its plans materialized.

In August 1998 the new coalition government headed by the Social Democratic Party (CSSD) declared that reform of the state administration would be one of its five key priorities. In its policy statement the Government set the year 2000 as the deadline for the approval of the Civil service Act. In March 1999 the government approved by its Resolution N 228/1999 the Concept of Public Administration Reform dealing almost entirely with territorial government reform.

In particular the new government announced its intention to:

- submit a number of new bills for implementation of the Constitutional Act on the establishment of a regional tier of government;
- adopt a Civil Service Act in order to depoliticise state administration, enhance its effectiveness and reduce its dependence on short-term political pressures;
- create a system of training of government employees;
- expand citizen's participation in governance;
- reinforce the economic powers of local and regional government.

In parallel, certain technological modernisation of the Czech state administration has been going on since mid 1990s. Government employees were provided with computers and respective training. Computer networks and on-line systems have been developed. In 1996 the Office for the State Information System was established.

However, it appears that various initiatives to "institutionalize" the Czech civil service at that time were effectively put on hold on their way to final approval by the Parliament and Government. For quite a long time the Czech Republic was ranked by the European Commission as an outsider among the Central and Eastern European countries (which made up the first group of those joining European Union) as far as administrative capacity of the state was concerned.

A Czech expert on public administration wrote in 1999: "When the former administrative regions were abolished, no effective structure to replace them was offered. As a result self-government got frozen at the village level. No civil service law was adopted. The state administration was largely politicized. Perhaps paradoxically some parts of government moved to even greater centralization. Procedures became less flexible and at the central level degenerated into mechanical job descriptions. Inefficient deconcentration of the ministries resulted in even more inflated regional bureaucracy so that in spite of Klaus's market rhetoric, at the end of his five-year term, the number of administrators at top levels had multiplied almost twofold from less than 100,000 to 170,000 (Potucek, 1999:1).

The post-communist transformation in the Czech Republic witnessed periods when "there was no body authorized by the government with overall responsibility for improving, modernizing and reforming public administration" (NTF 1998, Foreword:1).

6.3 NTF Report

A rare attempt to carry out a comprehensive analysis and assessment of the Czech state administration and thus to draw attention to its unsatisfactory status was undertaken by the National Training Foundation (Czech non-government non profit organisation) in 1998. The initiative was backed by international donor-organisations. Regretfully, the recommendations of the report, finalized in September 1998, remained largely unclaimed.

One of the report's strongest points was that it provided a substantial overview of constraints and problems facing Czech public administration. One can find in the report a remarkable statement "what is at issue in the Czech Republic, like in other European post-communist countries, are not partial reforms of territorial or functional administration, not a simple process of its gradual modernization through partial and minor changes but *a complete and principal qualitative transformation* of public administration" (NTF, 1998: 1-3-2).

The report continued by stating that "the public administration reform has been understood in the Czech Republic in a very narrow way as the setting-up of regional authorities and the adoption of the Civil service Act. The aims of the reform, which would secure the compatibility of our public administration with the Western level are much wider and deeper, they concern the very substance of public administration, its role, functions and means. Therefore, it will be vital to follow the vision of the desirable future development of the Czech society, economy and state. The decisive question is: what kind of state do we want and what role should this state play in the future development of our society and country?" (NTF, 1998:7-1-7).

In comparison to most other post-communist countries where fundamental reforms of civil services had been long part of the practical agenda "the Czech Republic was lagging obviously behind most of those countries… as they have already created the foundations of the legal regulation of civil service and modern personnel management, established or modernized training institutions for civil service and have bodies responsible for this area" (NTF, 1998: 6-1). Assessing previous attempts to develop and adopt reform programmes and civil service legislation in the Czech Republic the authors of the report pointed to the principal drawback – that promised and planned measures and actions remained only on paper.

The report also comprised well balanced recommendations concerning possible strategy of reform of the Czech state administration. It is regrettable that the NTF report was unknown even to some specialists of public administration in the Czech Republic.

6.4 Civil Service Reform Preparation and Implementation Stages

The analysis of the reform conceptual documents and statements made by the reform managers allows single out the following stages of the preparation and implementation of the Czech civil service reform.

I. REFORM PREPARATION 1992 – 2004

a) 1992 – 1997 this period was characterised by some rather fragmented actions resulting in particular in several drafts of civil service law being developed, submitted to government and parliament but each time rejected, also a number of normative acts dealing with various aspects of activities and position of government employees were adopted;

b) 1998 – 2000 the period is known for overall greater attention to the issue of public administration and civil service reforms, mainly as a result of growing EC pressure;
c) 2001 – 2003 during this period the Concept of Modernization of Central State Administration (Resolution 619 of 20th June 2001 of the Czech Government) was approved and Civil service Act adopted in April 2002;

II. REFORM IMPLEMENTATION 2004 – 2010

a) 2004 - 2007 Government Resolution N 237 of 17 March 2004 approved "Concept and main directions of reform and modernization of central state administration", implementation of the reform started the same year;
b) 2008 – 2010 the reform conceptual documents envisaged transfer from reform to modernization of the Czech civil service (2006-2007 finalization of functional review/audit; 2008 – 2010 implementation of the review conclusions).

6.5 Political Leadership and Reform

Throughout most of the 1990s the Czech political leadership was giving priority to large scale privatisation and maximum reduction of the role of the state in the economic and public spheres. In spite of much rhetoric interest of the ruling parties, their leadership in an in-depth transformation of the Czech state apparatus was rather superficial.

The views of V. Klaus, leader of the Civic Democrats and the Czech Prime Minister in 1993-1996, on this issue are of special interest as his party was in power for over six years after the disintegration of Czechoslovakia. At that time V. Klaus was supporting the idea that the forces of an unrestrained market would solve all the problems of the country's development including those of state bureaucracy.

It is noteworthy that on the eve of 2002 parliamentary elections V. Klaus promised that if his party wins it would negotiate "equal partnership" for Czechs in the EU. He was particularly critical of the European Commission's demand that the Czechs should introduce a law on the civil service without delay in order to pave the ground for a well functioning administrative system: "Individual states have varying civil service rules and regulations and the question whether or not we have a civil service law is entirely inconsequential with respect to our progress and preparedness to join the EU. The insistence of the European Union on this law shows a complete lack of comprehension with regard to the workings of this country" (Radio Prague, 12 May 2002).

The President of the Czech Republic V. Havel appears to be one of the few Czech politicians who was fully aware of the need for changes in the state administration. On numerous occasions he emphasized the significance of a stable, efficient and modern civil service and an appropriate legal and normative environment for its proper functioning.

Concerning issues of public administration in his country he stressed that: "The nervous system of the state is constituted by a network of self-government and state administration authorities. I consider it a crucial task for the coming period to undertake a reform of this system. This reform has been put off for several years and the delay has done this country a great deal of damage. It is only in recent days that you adopted the first piece of legislation that paves the way to it -- the constitutional act on regions. It will have to be followed by a series of other laws to develop the concept, as well as by a civil service act whose absence is beginning to pose tremendous difficulties..."

V. Havel continued: "I find it absurd that while we are building a market economy, many of us do not object to the fact that whole spheres of our public life -- state administration is one of them -- still bear the marks of the communist pattern of rule over the people, including a high degree of politicization. It is not true that reform of public administration will produce more bureaucracy and more bureaucrats. Unless it is disastrously mismanaged, it should achieve the very opposite" (Havel, 1997).

On another occasion V. Havel stated: "It is vital that Parliament approves a civil service law without delay in order to strengthen our administrative sector so that newly approved laws can be implemented in practice as soon as possible. It is likewise important to work on improving the "political culture" in this country, especially in view of the need to fight corruption and various irregularities such as the very high number of government contracts granted without tenders"(Radio Prague, 22 February 2000).

6.6 Civil Service Act

All the governments in Czechoslovakia and the Czech Republic that have existed since the year 1991 referred to the necessity of adopting a law on civil service in their programme declarations (Spidla, 2001). The Civil service Act was finally approved by the Parliament in April 2002 and this was actually the fourth attempt to elaborate the law and get it accepted by the legislators. The very first draft was submitted already in 1991 during the times of the Czechoslovak Federal Republic (Streckova, 1997).

Among prior attempts to draft the Act, most successful is considered to be that of 1993 (Draft Act N579). This draft was approved and supported by all Parliament committees but later "got lost somewhere in the government". In the opinion of some experts the weak points of this draft consisted in a lack of incentives for an improved performance of civil servants and higher costs of the proposed pension and salary schemes which were neither estimated nor funded. The Act was introducing "employment for life" civil service and did not envisage a transitional period. Enforcement of this draft would have meant that incumbents of positions would have become civil servants with absolute tenure over-night without any proper evaluation or any requirements of training and passing examinations. The main intention of the 1993 draft law seemed to be to reward those that had been employed in the administration for a long time, not to build a system for the future (Vonkova, 1996:120).

During the rule of the second coalition government (1996-1997) the Ministry of Labour and Social Affairs set up a working group which reassessed the previous draft and largely abandoned the principles of "employment for life" civil service. However, the government did not have time to review the prepared document as it resigned in November 1997.

The work over elaboration of the draft law was further pursued by the successive governments but neither civil servants nor citizens have been invited to be involved (Onisko, 2000:69). Similarly, assistance coming from international organisations was not always effectively used. For example, in 1995 SIGMA (OECD structural unit) provided a detailed commentary to the draft Civil service Act. But its conclusions were largely neglected (World Bank, 1999:7).

When the long-awaited, much-discussed and very controversial draft law was submitted by the governing Social Democrats to the Parliament for final reading and approval in April 2002 it was pushed through the lower House by just one vote! The opposition Civic Democrats were against it saying they saw no reason for civil servants to be elevated to what they called "special status". Besides in their view implementation of the law would require huge funding from the state budget. The Communists disagreed with the law's retention of the screening process aimed at preventing former high-ranking Communists and secret police collaborators from holding public office.

Generally, the process of civil service law preparation was characterised by an atmosphere of strict confidentiality, insufficient attention to prior resolution of conceptual issues and, at times, lack of "corporate memory" in the development of various proposals. According to Onisko "no alternative models of civil service systems had been considered before adopting the model which forms the basis of the Law" (Onisko, 2000:65).

The Act has 254 articles and three annexes. But only 15 articles and one annex were enforced upon promulgation (Vidlakova, 2006:2). By comparison the draft law from 1993 contained almost 400 pages (275 articles) plus attachments (Cebisova, 1996:136). By the year 2006 i.e. even before full enforcement of the Law it had been amended seven times! (Vidlakova, 2006:2). The Act was supposed to enter into force in 2004 and its full implementation was originally planned to occur by the end of 2006. However, full enforcement of the law was several times postponed with the latest deadline fixed for the year 2009.

In the opinion of OECD experts the text of the law had numerous inconsistencies, confused wording and structure. It was too large for such a legal act and overburdened with excessive details. This may lead to complications in interpretation and implementation problems.

The "EC Regular Report 2002 on Czech Republic's progress towards Accession" referred to the following major objectives of the Civil service Act:

- depoliticisation of personnel policy by the creation of a General Directorate for the Civil Service which will be responsible for the application of unified human resources management throughout the administration;
- rationalization of the number of civil servants posts and the funds earmarked for salaries in state administration;
- introduction of uniform professional standards through increased use of competitive exams for filling in vacancies and more transparent criteria for promotion;
- establishment of the Institute of Public Administration which will contribute to the creation of a comprehensive and upgraded system of civil servants' training;
- streamlining and standardizing the system of remuneration;

The law intended to limit political influence. A clear border-line between functions and positions staffed by political appointees and those occupied by career civil servants was to be established. The Act also envisaged higher pay and improved work conditions for civil servants in return for better skills, improved performance and greater independence from political influence

The law provided for the establishment of a HRM (Human Resources Management) Department in each ministry headed by a Personnel Director, who would be appointed by, and reporting to, the General Director for the Civil Service. The Personnel Director, responsible for selection and appraisal of all civil servants in a ministry would not be subordinated to the Minister or to anyone else in the ministry.

Provisions of the law envisaged precise definition of the authority of ministries, agencies and departments, as well as of individual civil servant's posts and required greater transparency from state administrative organs. Regular upgrading of qualifications and re-training was to become a civil servant's duty. The law envisaged that over the period 2004–2007 current personnel would apply for admission to a civil service "preparatory" stage and pass respective service examination in order to retain their posts.

The drafters of the civil service law seem to have been concentrating largely on employment conditions of civil servants. The law provided for considerable increase in remuneration of civil servants. The annual increase of the remuneration budget was calculated to be between 5.7 and 7.1 billion crowns (Debates in the Czech Parliament, 12 April 2002).

Overall the Civil service Act was a considerable step forward as it laid down legal foundation for the existence of a civil service in the Czech Republic. The new law also provided for introduction of better control over overall staff number and costs of the civil service. For that it envisaged preparation of annual personnel plans outlining the required number of staff, grades and salaries allocated to positions which should be managed from a central information database (SIGMA, 2003:22).

The authors of the law claimed that it was aiming at creating "an open system of civil service" with "no guarantee of life long employment" (Debates in the Czech Parliament, 12 April 2002). However, many experts expressed view that the proposed model comprises some obvious elements of the tenure system and therefore it can be classified minimum as mixed (SIGMA, 2003).

Analysis of the Act leads to the conclusion that it does not envisage an effective mechanism for ensuring improved management and performance. The law does not explicitly define such important issues as role and mission of civil service, civil servants' mobility and motivation, delegation of responsibility, and prerequisites for elimination of corruption. Measures to improve the quality of civil service personnel are generally limited. The duties of state administration and civil service in relation to citizens and other institutions are not well explained. The issue of accountability of civil servants remains largely unconsidered.

It appears that the increased expenditure caused by the law will mainly be used for rewarding officials retiring during the coming years and less for attracting adequately educated new candidates. Assessment carried out by OECD experts pointed out that the law granted "excessive benefits and privileges" to civil servants (SIGMA, 2003:4). Similar opinion was expressed during Parliamentary hearings that "the draft Law is not so much about duties but more about benefits" (Debates in the Czech Parliament, 12 April 2002).

The most feasible up-to-date implications of the Civil service Act is that by late 2007, i.e. 5 years after adoption of the law, most of its provisions had not been enforced yet.

6.7 Accession to EU and its Impact

As outlined above, in spite of numerous official declarations the Czech post-communist political leadership did not consider reform of the Czech state administration as its real priority for quite a long time. It is true that various plans and strategies of reform were developed both by government agencies and independent organisations. However intentions proclaimed in these documents remained mostly on paper. The only areas where essential changes (though not always for the better) did take place before 2000 were regional and territorial government.

By announcing its intention to join the European Union the Czech Republic undertook a commitment to improve among other things its state administrative structures and raise the efficiency of its civil service. Many observers expressed the view that it is only after criticism contained in particular in the EU annual reports on the status of preparation of the Czech Republic for accession that the Czech Government took practical steps to adopt a Civil service Act and launch civil service reform.

In fact accession of the Czech Republic to the EU offered in many aspects a unique chance of getting away with undesirable features of the "old" bureaucracy. New opportunities for creating a modern civil service in the Czech Republic had opened up. The question one may ask is whether the Czech Government used these opportunities.

It is interesting that one of the experts in the field used the expression "battering ram" to describe the extent of pressure from Brussels on the Czech authorities at that time. "Without the EU pressure, it is unclear whether any relevant reform would be passed in either two countries (Czech Republic and Slovak Republic - AK). The 'battering ram' function allowed to create credible exogenous pressure on individual politicians, political parties, government employees and stakeholders and break through a very complex web of interests, which would otherwise almost certainly veto the change" (Behlavy, 2002: 64).

Concern of the European Commission with the level and quality of public administration in accession countries was first made known publicly in Madrid in 1995 when the European Union outlined "adjustment of administrative structures" as a condition for accession. Some authors argue that only by 1997 the EU had developed a relatively clear picture of state administrations it wants its future member states to embrace (Scherpereel, 2002:2).

As far as the Czech Republic is concerned the European Commission was fairly consistent in emphasizing that the existence of efficient, accountable and legally protected civil service was one of the important prerequisites for accession. In its regular reports the EC stated that in the Czech Republic civil servants lacked the necessary security, were defenceless against politically motivated decisions and their income was considerably lower in comparison to those in the private sector. This resulted in low interest of the young generation in positions in government agencies, low motivation of civil servants, high turnover of personnel, tendency to seek higher income elsewhere, and relatively high level of corruption.

The experience of a number of CEE countries in the area of the improvement of public administration was discussed in May 1997 in Rotterdam at a session of the SIGMA (Support for Improvement in Governance and Management in Central and Eastern European Countries) programme funded jointly by OECD and EU. The document adopted in Rotterdam outlined some prerequisites for the success of the ongoing reform process. They comprised in particular:

- *Political will and support*: the most advantageous solution is that the authority responsible for the planning and overall implementation and coordination of the reform is as close to the Prime Minister and the Government as possible;
- *Strategic approach*: the reform must be broadly based and it must have an overall strategy; no matter how good the intentions may be, partial reforms without an overall strategy and without complying with joint principles and coordination cannot substitute the general reform;
- *Well-considered progress*: it is necessary to proceed in a well-considered, deliberate way.

The European Commission found it unfortunate that the Czech Republic did not have at that time a body responsible for designing and monitoring the implementation of administrative reform. It made a note that there was no civil service act in the country. It also observed that there was relatively widespread shortage of staff, especially of skilled staff in certain departments that are essential for applying the Community acquis.

There was a distinct difference in understanding the contents of the reform between the EU and Czech authorities. While Brussels was talking mainly about the reform of central level administration, Prague regarded the reform as being mostly about organisation of territorial government.

6.8 Assessment Made by the European Commission and OECD

Until the final decision on the membership of the Czech Republic in the European Union was taken the quality and status of its state administration had often been criticized by the European Commission.

A study ordered by the European Commission and published by EIPA (European Institute of Public Administration) in August 1997 evaluated the Czech Republic as follows: "Public administration reform in the Czech Republic is seriously lagging behind the approaches pursued in most of the neighbouring countries. There is a significant shortage of qualified personnel, institutional stability and procedural capacities. The emphasis on developing a pre-accession strategy should be used to engage in a serious attempt at reform. That ought to include not only further attempts at training (currently under way via the National Training Foundation), but also policies to functionally link legal approximation, administrative reform and EU integration. It might furthermore make sense to help in establishing a "focal point" for administrative reform to allow for better information and transparency concerning the various reform attempts. This way the very thin layer of qualified and knowledgeable personnel and the few programmes serving as "islands" within a still anti-administrative and anti-public environment ought to be broadened" (NTF 1998: 1-3-3).

In another European Commission's document "Agenda 2000" commenting on the request of the Czech Republic for accession to the European Union the following assessment of the administrative capacity of the Czech Republic was made: "Since 1990, the governments of Czechoslovakia and later of the Czech Republic have given low priority to the necessary reforms and modernization of public administration. There are no signs of any changes in this attitude. The non-existence of a more significant or deliberate plan for the modernization of public administration represents the only serious reason for uneasiness in this area. Measures adopted are completely insufficient with respect to the significance of problems that necessitate a solution ... it will be necessary to start and keep in operation an extensive reform process if the Czech Republic is to introduce public administration of such general quality, level of knowledge, motivation and flexibility which will be necessary for the path of the country towards further economic and social development and its membership in the European Union."

Starting from July 1997 the European Commission opinions on the state of readiness of the candidate countries for EU membership used administrative capacity as a separate criterion (Verheijen, 2000:16). Since 1999 the evaluation of the administrative capacity was based on the regular SIGMA (OECD public management programme) baseline assessments, which focused on six core areas (ibid):

- Legal status of civil servants;
- Legality, responsibility and accountability of civil servants;
- Impartiality and integrity of civil servants;
- Efficiency in management of civil servants and in control of staffing;
- Professionalism and stability of civil servants;
- Development of civil service capacities in the area of European Integration.

It should be noted that the level of criticism by the European Commission and OECD of the Czech Republic's advance in the modernisation of its state administration did not ease at least until the year 2004. "With regard to civil service reform, the Czech Republic shows a disappointing track record as politicians, with some outstanding exceptions, have considered civil service reform unnecessary.... Given this state of affairs, it can be concluded that the Czech Republic has passed a law on civil service only to comply with the formal requirements posed by the European Commission's Regular Reports since 1998 and not because there is a genuine political commitment to make the civil service more professional and less dependent on political patronage" (SIGMA, 2003:1).

6.9 Civil Service Reform Preparation and Implementation Milestones

June 1992 - the Law 'On Conflict of Interest' (Act 238/1992) adopted, applies only to high level "public functionaries", ministers and heads of agencies;

October 1992 - the Office for Legislation and Public Administration of the Czech Republic (OLPA) established;

August 1993 - the Government approved basic principles of the Civil service Act;

November 1996 - the Government abolished the OLPA;

December 1997 - the Czech Parliament adopted a Constitutional Act on the Establishment of Higher Territorial Self-Governing Units in the regions (regarded at that time as major trend of public administration reform);

March 1998 - the Government assigned the role of central authority for public administration reform (including coordination of activities of other government and state bodies) to the Ministry of Interior.

August 1998 - the Government issued a Policy Statement which comprised commitments in the area of public administration reform (in particular referring to the necessity of adopting a civil service law) following EC Opinion 1997 on the status of preparation of the Czech Republic for accession;

May 1999 - the Law 'On Free Access to Information' (Act 106/1999) was enacted by the Parliament of the Czech Republic and entered into force on 1 January 2000;

May 1999 - the Government adopted Resolution (No. 511) on the Strategy of Public Administration Reform;

June 1999 - the Government adopted Resolution (No. 601) approving the Concept of Training of Civil Servants;

July 1999 - principles of public administration reform approved by the Resolution of the Government (No. 716);

December 1999 - establishment of the Institution of Ombudsman (Act 349/1999);

December 2000 - draft Civil service Act submitted to the Czech Parliament;

March 2001 - Code of Ethics of state administration employees approved by the Government;

July 2001 - establishment of the Institute of Public Administration;

July 2001 - Government Resolution (No.619) approved the Concept of Modernization of Central State Administration Considering the Status of Civil Servants and Structure of Administration Authorities;

April 2002 - the Parliament adopted a Civil service Act (No. 218/2002) on 26 April 2002;

June 2002 - The General Directorate of Civil Service created within the Office of the Government on 15 June 2002;

August 2002 - Programme declaration of the Government formally announced the reform of central state administration as one of its priorities;

January 2003 - Supreme Administrative Court established;

March 2004 - Government Resolution (No. 237) of 17 March 2004 approved "Concept and main directions of reform and modernization of central state administration". The resolution launched implementation of the reform of central state administration in the Czech Republic.

July 2005 - the document "Further Advance of Improvement of Central State Administration in the Years 2005–2006" which specified the main directions of the reform of central state administration was submitted for consideration of the Government;

November 2006 - the Government adopted a Resolution (No.1232) to transfer management of civil service reform back to the Ministry of Interior;

August 2007 – the Czech Government decided at its meeting on 13 August to abolish the Civil service Act.

6.10 Concept of Public Administration Reform

In course of developing various concepts of administrative and civil service reform the objectives, tasks and contents of the reform underwent certain and sometimes considerable evolution and change. As one may presuppose there existed and still persist in the Czech Republic differing, in some cases polar, views and opinions as to the overall strategy, directions and specific contents of the reform. Nonetheless, implementation of civil service reform (reform of central state administration) launched in 2004 was largely based on the provisions and guidelines contained in the document "Concept and main directions of reform and modernization of central state administration" approved by the Government Resolution No. 237 of 17 March 2004. This document deserves special attention and will be reviewed below.

Broader view on the essence and scope of public administration reform which goes far beyond just reform of regional and territorial government started to acquire prominence in the Czech Republic in the late 1990s. Several documents on the strategy of public administration reform were drafted at that time. The Ministry of Interior used them for elaborating a new conceptual approach for public administration reform. A working group composed of respective Deputy Ministers and heads of some central state agencies was created in September 1998, headed by the Deputy Minister of Interior who was in charge of public administration reform. This group performed the function of a co-ordinating and advisory body. A special official was appointed in each ministry as a reform coordinator. Every three months the group prepared a quarterly report which was signed by the Minister of Interior and then submitted to the government. As a result of the activities of this group and of a section in the Ministry of Interior a draft Concept of Public Administration Reform in the Czech Republic was elaborated.

The Concept of Public Administration Reform and the Concept of Training of Public Administration officials were approved by the Czech Government in 1999

- Resolution No. 228 of 30 March 1999 and Resolution No. 601 of 16 June 1999 respectively (see National programme for preparation of the Czech Republic for EU membership, June 2001 point 1.1.1). The first document stated that public administration reform in the Czech Republic "consists of three fields: 1.Territorial structure reform of public administration; 2. Reform of public administration at the central level; 3. Reform of the functioning of public administration." Such differentiation seems largely illogical since the two fields are determined on the basis of their place in the overall structure of governance (central and territorial levels) and the third field on the basis of subject matter (activities).

The Government decided about further progress of the reform at its session on 26 May 1999 by adopting Resolution No. 511 on the Strategy of Public Administration Reform. The resolution envisaged that for public administration reform a whole package of conceptual documents should be drafted. The list of concepts comprised those on:

a) preparation and training of officials in public administration;
b) informatics in public administration;
c) improving the efficiency of public services;
d) improving quality management in public administration;
e) improving the effectiveness of public control.

Subsequent positive developments in this area included: approval of the principles of civil service reform by the Government Resolution No. 716 of 14 July 1999; creation of a governmental board to deal with administrative reform issues composed of six ministers presided by the Prime Minister and of a working committee in which deputy ministers of all interested ministries took part (Onisko, 2000:5).

As we can see the list of conceptual documents approved by the Czech Government and devoted to public administration reform is large (we have mentioned not all of them by far) and rather confusing. We should bear in mind that apart from the Civil service Act (adopted in April 2002) there exist two principal documents, approved by the Czech Government, which largely determined the dimensions, contents and general time frame of civil service reform:

1) Concept of Modernization of Central State Administration Considering the Status of Civil Servants and Structure of Administration Authorities - (Government Resolution No. 619 of 20 June 2001);

2) "Concept and Main Directions of Reform and Modernization of Central State Administration" (Government Resolution No. 237 of 17 March 2004).

The first concept dealt with several issues including territorial reform, reform of central administration and overall reform of governance in the Czech Republic. The second concept was devoted entirely to the reform of central state administration. In defining specific ways and instruments to implement reform the first concept referred to the experience of the European Union member states in reforming their administrative systems over the past several decades "The targets of the modernisation of the central state administration in the Czech Republic are similar to those for which also other advanced countries are striving approximately during the last twenty years, in particular the member countries of the European Union and, in the bigger picture, also the member countries of Organisation for Economic Co-operation and Development (OECD)" (Concept, 2001).

The same concept singled out the following main lessons that could be drawn from foreign and Czech experience:

- public administration reform must become a government priority long-term in character; the reform should be headed by the Prime Minister;
- public administration reform must become a public matter; it must acquire public support both on the level of citizens and on that of public administration officials and staff;
- the concept of pubic administration and the methods of its implementation must be based on political consensus between the government and the opposition which is possible only if the institutional structure, processes and technology of public administration are considered above party level;
- public administration reform must absorb the preparation for the entry to the European Union in its very concept, i.e. in its institutional structure, processes and technology, the concept of European public administration in the European administrative space.

The Programme declaration of the Government from 20 August 2002 defined the objective of the reform of central administration in the following manner: "Greater efficiency, rationalization and horizontal coordination of state administration activities by means of wider application of modern management and IT technologies with the aim of raising overall quality of administrative performance for the benefit of citizens" (Radio Prague, 21 August 2002).

6.11 Reform Programme.

The objectives and tasks of civil service reform as well as programme of its implementation were outlined in the above referred document "Concept and Main Directions of Reform and Modernization of Central State Administration" (Government Resolution No. 237 of 17 March 2004). Various aspects of reform implementation were specified further in the document "Further Advance of Improvement of Central State Administration in the years 2005–2006" submitted to the Government for its consideration in July 2005.

The main reform goal is defined as "To ensure that state administration bodies implement better the provisions of the law and serve government and citizens more effectively. Civil service should be more oriented towards quality of services and results".

Vision of the reform constituted "Flexible and well functioning central state administration responding better to global challenges and cross cutting issues". The overall approach to the reform was: "to ensure transfer from offer driven to demand driven model of central state administration" (Concept, 2004).

The Concept singled out five principal directions of reform:

A. rationalization of processes in central state administration;
B. improvement of management in central government;
C. raising quality in central government;
D. improving civil service in central government;
E. rationalization of public administration financing.

Implementation of the civil service reform was stated to be based on project management principles. In other words, the reform programme would be executed in the form of projects, each targeting a specific reform objective. A summary of the contents of the reform directions/projects is given below, whereas more detailed description of the projects and the status of their implementation by mid 2006 are provided in Annex 2 to this study:

A. Rationalization of processes in central state administration

- *to define mission statements of central government agencies and institutions and determine their mid term priorities;*
- *to implement audit of processes in central government agencies;*
- *to define standard rules with reference to support and operational services that should be provided to central government authority;*
- *to define the rules for the "new" central administration (including the old ones which proved to be effective);*

B. Improvement of management in central government

- *to define the rules of effective coordination among central state administration agencies and bodies;*
- *to develop a training system that will ensure application by top management of modern managerial techniques;*
- *to develop an effective mechanism of coordination between central state administration and territorial public administration;*
- *to develop a system of information sharing among central state administration as well as that of keeping "institution memory";*

C. Raising quality in central government

- *to train management of central state administration in quality management issues;*
- *to implement Regulatory Impact Assessment (RIA) in the legislation process;*
- *to promote e-government tools and practices;*

D. Improving civil service in central government

- *to implement provisions of the Civil service Act;*
  (namely, creating personnel departments in each government ministry and agency; transforming the Institute of Public Administration into a state contributory organization; preparing organizational structure of administrative authorities for the year 2005; issuing service regulation by the Director General of Civil Service; developing guidelines for training civil servants; drafting plan for supervisory activities of the Directorate General of the Civil Service for the year 2005; elaborating principles for appraisal of civil servants;

E. Rationalization of public administration financing

- *to prepare top management in central administration for wide economic reforms;*
- *to define the rules for the use of public and private resources in public investments;*
- *to enhance systematic control and strengthen its relation to higher performance.*

6.12 Management of Preparation and Implementation of Civil Service Reform

At different stages of preparation and implementation of public administration and civil service reform in the Czech Republic overall responsibility rested with various organizations. Often this responsibility was fragmented, overlapping or was entirely lacking.

A special body responsible primarily for the modernisation of Czech state administration was established in the Czech Republic in October 1992 and existed until November 1996. The body was called the Office for Legislation and Public Administration (OLPA). It reported to a Deputy Prime Minister. The Office comprised a special public administration division which served mainly as an advisory body and a point for elaboration of reform proposals. OLPA took an active part in the development of conceptual approaches to public administration and civil service reforms though few of them were resurrected for practical work later. The Office was dissolved in 1996 under pretext of the necessity to decentralise responsibility for reform to each individual agency.

Following dissolution of OLPA part of responsibility for preparing public administration and civil service reforms was divided between the Ministry of Labour and Social Affairs (MLSA) and the Ministry of Interior. In practice MLSA was dealing mainly with developing relevant legislation including draft civil service law. The Ministry of Interior was responsible more for overall coordination and elaboration of the reform conceptual documents. In 1998 the Czech Government formally entrusted the Ministry of Interior with drafting of the concept of public administration reform and co-ordination of its implementation.

Starting from 2002 reform management was transferred to the Office of the Government - in particular to the Department of the Regulatory Reform and Central State Administration Reform. The General Directorate of Civil Service (set up in 2002) was also involved in the management of reform activities. Operational management appears to have been ensured by the Reform Steering Committee composed of Deputy Ministers and Heads of large structural units of respective ministries. Overall supervision was exercised by the Prime Minister.

Implementation of civil service reform (reform of central state administration) was formally launched in March 2004 when the Czech Government approved the document "Concept and Main Directions of Reform and Modernization of Central Public Administration". This document mapped out the main directions of civil service reform executed in the form of projects (see above). It defined the project outputs, deadlines for presenting results and those organizations and persons responsible for their implementation.

By adopting this document the Czech Government ended the reform preparation phase and launched the implementation phase. Application of project management principles in reform implementation meant that major components of the reform had been partitioned into separate projects each targeting a specific reform objective. Each project was implemented by a separate project team headed by a project manager. Each ministry and other executive agencies appointed so called "managers of public administration reform" in all large structural units to ensure greater coordination of reform activities.

In terms of reform implementation tools the Czech authorities gave exclusive preference to the development and adoption of legislation. Other tools such as training and best practice diffusion were applied occasionally. The importance of fiscal incentive instruments, networking and rotation was largely disregarded.

Another quite essential dimension of the good quality reform management was notably absent. Full enforcement of the law which provided the legal ground for all the changes and innovations in the Czech state administration, has been postponed four times. The latest deadline was fixed for the beginning of 2009.

6.13 Reform Diffusion and External Assistance

In designing a strategy of civil service and public administration reforms the Czech Government and its experts were resorting to external assistance coming within the framework of bilateral agreements with predominantly EU member countries and contributions made by international organisations, mainly the European Union and OECD.

For instance, the years 2002–2003 saw realisation of the twinning project *Modernisation of the Central state administration in the Czech Republic* financed by the EU PHARE programme. The general aim of the project, led by Finish and French experts, was to provide recommendations on raising efficiency and performance of the Czech state administration through application of performance management and performance budgeting, total quality management, wide application of IT technologies and other tools and methods. During project implementation working groups on the modernization of central state administration were created in most ministries and agencies. This was considered necessary to insure proper coordination and interface between various government institutions in the process of reform.

The recommendations made by the project team, in particular concerning various aspects of establishment of the General Directorate of Civil Service, were reflected in the Government's "Concept and Main Directions of Reform and Modernization of Central State Administration" (Government Resolution No. 237 of 17 March 2004) (Jalovecka, 2004:8).

Another example refers to an evaluation of regulatory mechanisms and framework in the Czech Republic carried out by the OECD experts in June 2001. This work was assessed by the Czech Government as a fundamental contribution to the public administration reform in the country.

Co-ordination of external assistance funded by PHARE programme was carried out by the Foreign Assistance Centre in the Czech Ministry of Finance but, in view of some experts, was not always appropriate. "The use of external assistance for public service development has been scattered over individual ministries and other administration authorities without any co-ordination" (Vidlakova, 1999 point 8-10).

There appear to be no grounds for asserting that a single civil service model of a particular country or group of countries was used by the Czech authorities in shaping their perceptions as to what kind of civil service should exist in the Czech Republic.

Various views have been expressed concerning the absorption capacity of the recipient side and the value of rendered assistance. Although the overall positive effect of the external assistance can be hardly denied there was criticism as well. "Western specialists did not really understand the organization of Central and Eastern European societies.... It is rather surprising, for instance, that these Western specialists did not realize from the beginning that what these societies primarily need is an efficient and impartial civil service on the one hand, and the rule of law, on the other." (Culik, 1999).

The NTF report made the following remark. "A number of studies have been made for the support of the reform in Central and East European countries and numerous exchanges of experience have taken place. So far the utilization of these opportunities has been weak as foreign contacts have been limited to a very narrow group of specialists with the necessary knowledge of languages, there is no national base for the study and research of public administration and for the analysis of policies and strategic studies in this area (whether at government or university level). As a consequence, significant documents and studies such as OECD recommendations for deregulation, the Rotterdam document of the SIGMA programme, the EIPA study (Maastricht 1997) and many others are not only unfamiliar to our authorities (not even mentioning the public), but they are not available within one place and, moreover, even the authors of various reform proposals often do not know them either". (NTF, 1998: 7-1-13).

6.14 Assessment of Reform Implementation and Current Results

Assessment of the Czech civil service reform, its programme and extent of up to date implementation has been carried out against the declared outputs, deliverables and time frame (contained in the concept of reform adopted in March 2004), against provisions of the Civil service Act (2002), as well as against the recommendations and criteria developed by the National Training Foundation (Report 1998) and the OECD/European Commission (SIGMA, 2003).

On the evidently positive side it should be acknowledged that the Czech Republic has finally "acquired" its civil service as the Civil service Act has formally "legalized" this institution in the Czech Republic and determined the general status of civil servants.

The law introduced a number of far reaching (at least in the Czech context) innovations e.g. mandatory competition for vacancies and promotion, formal differentiation of political and career appointments, formation of a single HR management center for the entire service, cancellation (at least formal) of guaranteed life long service, to name just a few. However since most of the law articles "wait" to be enforced it is difficult to say how they would be implemented in practice.

Among other positive developments one should mention creation of a sound reform strategy and programme based on project management principles. The Civil service Act (to a lesser degree) and the reform programme (to a greater degree) are congruous with modern trends in public management development and civil service reforms and result from the respective experience of other countries, above all EU member states. Establishment of the Institute of Public Administration in July 2001 was also an obvious step forward.

There has been much wider involvement of civil servants in various forms of re-training and up-grading of qualifications. Appropriate training courses focused particularly on the application of the Common Assessment Framework (CAF), which is considered to be an effective tool in implementing changes in the management of state administration (Vidlakova 2006:4).

Regretfully, the drafters of the civil service law seem to have been concentrating largely on improving employment conditions of civil servants. This is in contrast to recommendations developed by the NTF and OECD/EC which insisted on a comprehensive approach proceeding above all from the conceptual understanding of the role and mission of civil service in the society.

Furthermore, permanent postponement of the full enforcement of the Civil service Act (the latest in order Government decision 531/2006 of 7 December 2006 moved the deadline to 1 January 2009) made this legal document quite unique in terms of the time frame of its validation – almost seven years (2002 to 2008)! - and raised serious doubts whether it would be ever implemented.

Quite disappointing remains the fact that by mid 2007 (5 years after adoption of the law!) the Head of the General Directorate of Civil Service had not been appointed yet. "The fate of the General Directorate, which was established in conjunction with the Act, does not seem to portend a positive future for the regime foreseen by the law; established in 2002 with 65 employees, later budgetary cuts targeted at human resources caused the Directorate to shrink to only 22 employees by spring 2005" (Landová, 2006:60).

Comparison of the recommendations contained in the SIGMA report (2003) with the contents of the reform programme and extent of its implementation reveals that so far only few recommendations have been taken account of. These include for example the need to place reform implementation issues under highest possible authority, expanding a range of training programmes for central government officials, and compulsory competition in appointments and promotion. On the other hand, at best insufficient attention has been paid to the development of accountability and anti-corruption mechanisms, improving transparency, speeding up enforcement of the new law, reducing managerial discretion in determining individual salaries, etc. Many of these issues remain unattended as most provisions of the new law have not been enforced yet.

With regard to implementation of the reform programme adopted in 2004 (Concept and Main Directions of Reform…) the majority of projects were launched to schedule and three sub-projects were terminated by mid 2006 (e.g. projects A1 *Identification of missions and functions of government agencies)*. However it is not always easy to evaluate extent of success of these projects since in many cases the expected results (outputs) had not been sufficiently detailed and proper evaluation or feedback was in most cases absent.

The government has issued resolutions approving a number of documents prepared by the project teams: No. 420/2005 'On introducing methods of regulatory impact assessment (RIA) in relation with businesses'; No. 421/2005 'On reducing administrative burden on entrepreneurs'; No. 430/2005 'Analysis of coordination mechanisms between central and territorial authorities'; and No. 738/2005 'Analysis of the system of control in central government'. At the same time there were serious setbacks where certain projects have not been launched at all. For example, project A4 (*Reorganization of central government),* appears to have been cancelled all together. The explanation given - the failure of the project A2 (*Description and analysis of processes*) the results of which should have been used in the project A4. (Further Advance..., 2005: 4).

Project A2 has largely failed "as *process of audit* was carried out only in the Ministry of Finance" (ibid). No reason was given (at least in the report on progress) why only one ministry was subjected to this audit procedure. Later it was reported that in spring 2005 the Government declared process audits no longer compulsory and that they could be exercised only on request of a particular ministry or agency. Again one may ponder the causes of such a decision. In our view such a selective approach marks a retreat from general principles of reform implementation.

Similar and a somewhat mysterious saga befell another quite important project D1 (*Implementation of civil service law*). Since late 2005 reference to this project dropped from reports on reform implementation. Several months later one of the documents prepared for the Government (attachment 1 to 'Further advance...') referred to a proposal to change the title of this project to *Raising quality of HR management system* (including proposals to modernize Civil service Act). To what extent the original objectives and tasks of this project have been retained remains unclear.

Some projects were cancelled on request of separate ministries e.g. project E2 on request of the Ministry of Finance. In a number of cases the time frame of the projects' implementation was extended. Justification of such measures was not always convincing. Overall these examples provide evidence of at times serious difficulties experienced by the project teams and of covert constraints and obstacles they encountered in advancing reform implementation.

Noteworthy information about the internal mechanism of reform implementation and encountered problems could be drawn from the document entitled "Further Advance of Improvement of Central State Administration in the years 2005–2006" submitted to the Government for its consideration in July 2005.

In particular, this document pointed out that reform implementation revealed serious drawbacks in distribution of information about reform. "There still exists overall very low knowledge about the start of reform implementation and its goals. Not only wider public but also many civil servants and the highest management of the central government institutions are unaware of the reform objectives and its main directions" (Further advance..., 2005:6).

Another concern expressed in the document referred to frequent changes of the government as a source of great difficulties and complications in project implementation (Further advance..., 2005:6). For example resignation of the Cabinet in summer 2004 and subsequent negotiations on the creation of the coalition Government resulted in replacement of the top management of many central government agencies. This quite negatively affected the work of reform projects teams.

Also reference was made to insufficient fulfilment of government resolutions. For instance, the Cabinet's requirement to all central government institutions to submit (by 30 June 2004) an analysis of all the implemented and planned projects in the area of application of ICT in their public relations activities was either completely ignored (e.g. Ministry of Justice, Ministry of Environment, Ministry of Defence, Central Statistical Office) or was implemented very superficially (Ministry of Agriculture, Ministry of Health, and Ministry of Interior) (Further Advance… 2005:6).

Besides, the reform has been often perceived as pure reorganisation, or as one time process, which can be resolved by a single political decision or legislative act. It also appears that issues of transparency, accountability, public consultations have been so far generally disregarded by the reform managers.

A distinct problem emerged regarding insufficient coordination of the project activities and often low respect for the agreements concerning overall management of the reform. In practice the Reform Steering Committee has not always been informed (as it should be) about the measures and activities undertaken by various ministries and agencies in the areas of project teams' intervention. This led to duplication of efforts and wastage of resources (ibid).

The document referred to numerous cases of "passive resistance" on the part of civil servants. This could be explained by "general shortage of information about reform and absence of a clear cut signal from either political leadership or top management in central government that implementation of civil service reform is an obvious priority of the Government" (Further advance…, 2005:7).

However most significant developments in terms of reform advance and results took place in late 2006. By approving Resolution No. 1232 of 25 October 2006 the Government decided to transfer coordination of the civil service reform from the Office of the Government back to the Ministry of Interior. The contents of the reform were seriously scaled down with only three projects out of the original list of seventeen remaining for further implementation. Simultaneously, the Institute of Public Administration, which was part of the structure of the Office of the Government, was also handed over to the Ministry of Interior. By early 2007 information about advance of implementation of civil service reform disappeared from the web site of the Czech Government.

This decision of the Czech Government looked particularly odd since before the reform was launched in 2004 its preparation had been managed by the Ministry of Interior and continuation of this practice was declared by the Czech authorities to be unacceptable. It was alleged that any single ministry was not in an appropriate position to lead reform efforts targeted at the entire government structure.

The culminating point of the long term "saga" of civil service reform preparation and implementation in the Czech Republic (though expected by "shrewd and experienced" observers) happened on the 13 August 2007. On that day the Czech Government adopted a resolution cancelling the Civil service Act and announcing that a new law on civil service would be drafted. Explanation given was neither amplified nor convincing – "the Act granted too many privileges to civil servants and did not sufficiently motivate good performance". It remained unclear, however, why over the five years (a quite sufficient time period) that elapsed since adoption of the Act there had not been improvement of this legal document.

Persistent unwillingness of the Czech authorities to enforce the Civil service Act, complemented by drastic reduction of the number and scope of reform projects and by the transfer of the reform management back to the Ministry of Interior leave us only one option i.e. to conclude that by late 2006 implementation of civil service reform in the Czech Republic had been put on hold. In this connection is a rather remarkable comment made by a Czech expert in the field: "While we cannot discount the burden of the negative connotations of civil servants in the past, it must be said that the improvements – not only in the image but also in the actual state of affairs – were hindered by a constant lack of true advocates for the civil service within the Czech political elite." (Landova, 2004:57).

6.15 Conclusions

Throughout most of the post-communist period reform of the state administration in the Czech Republic was largely seen as a task of "reducing" the state or leaving it "undisturbed" rather than "rebuilding" the state. In spite of at times lavish rhetoric the Czech post-communist political leadership did not consider reform of the Czech civil service as their real priority throughout most of the 1990s.

During the 1990s the internal factors pushing for civil service reform were weak and sporadic. They were mainly represented, on the one side by opposition parties and groupings which used the issue of corrupt and inefficient bureaucracy to blame their political opponents, and on the other side by non-governmental and civil society organisations concerned with the issues of adequate development of democratic institutions in the country. The latter have not been strong enough to make the government sustain its commitment to the civil service reform at that time.

During the 1990s public administration reform in the Czech Republic was viewed almost exclusively as reform of the regional and local government whereas civil service reform proper was largely confined to the adoption of civil service law. Although various drafts of civil service law and strategies of civil service reform were developed (as shown above) by government agencies and independent organisations, intentions proclaimed in these documents remained mostly on paper. The only areas where essential changes did take place before 2000 were the organisation of territorial government.

Insufficient quality of the Czech civil service has been a permanent cause of criticism by the European Commission. The process of accession of the Czech Republic to the EU offered in many aspects a unique opportunity for cardinal improvement of its state administration. Under strong pressure applied by the European Commission in the late 1990s the Czech Government finally undertook practical measures aimed at developing the necessary legal framework and launching civil service reform.

Following approval of the strategy of reform of central state administration in June 2001 and adoption by the Czech Parliament of the Civil service Act in April 2002 the Czech Government approved in March 2004 a programme of civil service reform which was effectively launched the same year.

Raising efficiency of the civil service activities was declared the principal reform objective. The authors of the civil service law and reform conceptual documents declared that they intend to build up an "open" position based civil service system. The reform programme comprised the following major tasks :

a) carrying out functional analysis; b) reorganising the Government; c) improving coordination and communication; d) defining standards and regulations; e) introducing modern management technologies; f) enforcing civil service law; g) promoting e-government practices; k) improving monitoring and control; l) enhancing training of civil servants; m) developing quality management; and n) improving financial management.

Implementation of the reform programme was based on the application of project management principles whereas adoption and enforcement of the respective legislation turned out to be the main reform implementation tool. In designing reform conceptual documents the Czech experts drew largely from international experience, above all that from EU member countries. There seem to be no grounds to assert that any single civil service model was borrowed for replication under Czech conditions.

Available information allows conclusion that opposition to civil service reform in the Czech Republic is considerable and has taken different forms. The current results of the reform implementation are rather modest and, as recent developments show, are in a way disappointing. The Civil service Act was adopted by the Parliament in April 2002 but its full enforcement has been regularly postponed since then. The outcomes of the reform implementation have been confined so far mainly to the development of methodological instructions and manuals on how to introduce innovations and improvements, establishment of the Institute of Public Administration, wider scale of civil servant re-training and application of information technologies.

By early 2007 the Czech Government had not yet appointed the Head of the General Directorate of Civil Service (envisaged by the law), considerably reduced and modified the scope of reform activities, transferred the function of reform management back to the Ministry of Interior, and eventually abolished the Civil service Act in August 2007.

The only conclusion that can be drawn from these developments is that the advance of civil reform in the Czech Republic by that time (2007) had been halted. It also means that the adoption of civil service law and the start of civil service reform in the Czech Republic were realized mainly for the sake of meeting external requirements rather than due to internal conviction or imperative.

CHAPTER VII
COMPARATIVE ANALYSIS OF CIVIL SERVICE SYSTEMS IN THE RUSSIAN FEDERATION AND THE CZECH REPUBLIC: IS THERE MORE DIFFERENCES OR SIMILARITIES?

7.1 Introduction

This chapter introduces the reader to the results of a comparison of prime features of the Russian and Czech civil service systems carried out in line with the Protocol (guidelines) for comparative studies of national civil service systems developed by experts from Leiden and Indiana Universities specifically for comparative research of administrative systems.

In accordance with this methodological framework the principal characteristics of the two civil service systems are grouped into clusters – historical development, internal labour market, representativeness, politicization, public opinion, and reform and diffusion. Within every cluster we examine its key elements, presented in most cases under respective sub-titles.

The cluster 'historical development of the nation and civil service' was reviewed at the beginning of each country specific chapters (Chapters III and V) and therefore is not present in this chapter.

In view of its direct relevance to the objective of our research the cluster 'reform and diffusion' will be considered in the next chapter devoted to comparison of civil service reforms proper.

The analysis and comparison of certain characteristics of the civil service systems in the Russian Federation and the Czech Republic gives us an opportunity to see a comprehensive picture of the problems experienced by each system, its status and reform capacity and, no less important, to single out similarities and differences in the two systems.

7.2 Czech and Russian Civil Services: Selected Features According to the Protocol for Comparative Studies.

In the analysis pertaining to each of the major topics (as defined in the Protocol) we make a distinction between the situation in the Czech civil service before and after the adoption of the Civil service Act (2002). Accordingly, similar distinction was made between the situation in the Russian civil service before and after the start of civil service reform in 2003.

Once again it should be emphasized that there exist definite shortage or actual absence of reliable statistical data and information on various aspects of activities of the Czech civil service. This was in particular emphasized in the report prepared by the National Training Foundation which pointed to a lack of plausible and usable statistical data, relevant databases, publicly accessible inspection reports etc., on the status of the Czech public administration (NTF, 1998, foreword). A similar statement was made also in the World Bank report (World Bank, 1999:3).

The situation with regard to statistics concerning the Russian civil service is overall better though considerable "black holes" or "omissions" in availability of information exist here as well. Therefore, wherever appropriate, absence of respective data was substituted by expert opinions and assessments.

For the purpose of objective comparison we tried to use, wherever possible, information and data from the years 1998-2002 (a decade after the start of political and economic transformation) and 2003-2005 (the start of civil service reforms).

7.3 Internal Labour Market

The Internal Labour Market is defined as the "rules pertaining to job definition or classification, deployment, job security and membership, and reward structures" (Wise, 1996). One of our objectives consists of tracing variations and similarities between the two civil service systems in the way human resources are used and rewarded. The importance of studying the rules governing entry to civil service and various aspects of human resource management can be hardly overestimated as it allows us to identify and determine the extent of democratization and social equity of a civil service as well as the factors influencing its efficiency. By "rules" we mean not only formal but also informal rules. Informal rules may be quite important in understanding key elements of the organisational culture and bureaucratic tradition and may provide explanation of the direction and speed of reform implementation.

In exploring various aspects of internal labour market we have paid priority attention to:

a) job definition and classification system;
b) deployment rules;
c) job security and membership rules;
e) reward mechanisms and practices.

7.3.1 Job Definition and Classification System - the Czech Republic

According to World Bank estimates *the number of officials* in the Czech state administration at central and regional levels amounted to approximately 154,000 in 1998 (Nunberg, 2000:122). The respective figure for the year 2004 was 186,000 (Debates in the Czech Parliament, 12 April 2002). According to Zuzana Rusňáková from the Office of the Czech Government there were about 108,000 officials working for the central government in Prague and in the regions in 2004 (Rusňáková, 2004).

Taking into account total employment in central budgetary and subsidized organizations this number reaches around 508,000 employees and constitutes 7.2% of the population (World Bank, 1999:122). The increase in the number of public administration officials is estimated to be over 1.8 times for the period 1992–2002 (World Bank, 2006:122).

In terms of *job classification* we should note that until 2002 there has been no civil service legislation, no distinct group of employees called "civil servants" and no civil service *de jure* in the Czech Republic. Respectively the scope and the nature of civil service were not formally defined.

Still, *de facto* there existed a category of state administration officials, although their status and conditions of employment were regulated largely by the Labour Code. Article 73 of the Labour Code refers to a group of government employees who may be considered, de facto, civil service staff. As we can see the status of these employees was not different from the status of other employees in the public sector. Until recently civil servants were employed by individual ministries and there did not exist a strategy of personnel deployment across the whole state administration.

Table 7.1. Public Sector Employment in Percent of Overall Population

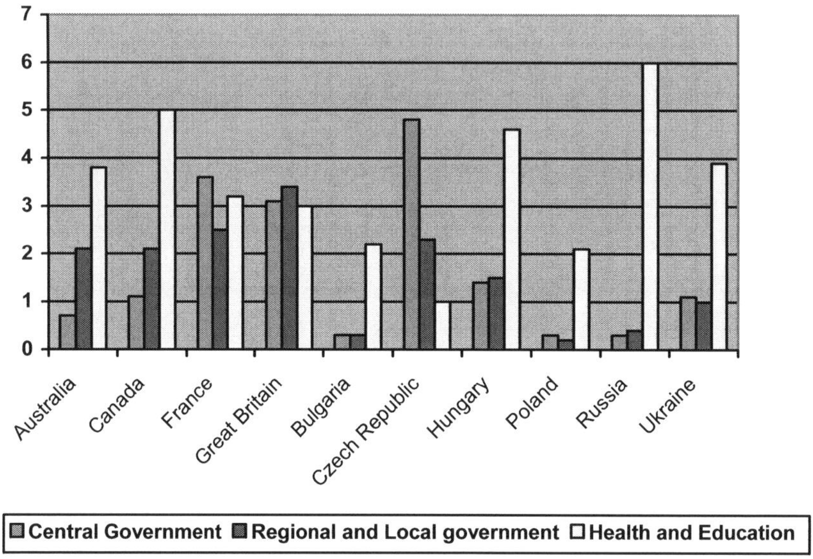

Source: World Bank staff, Moscow, 2005

Table 7.1 shows that the Czech Republic and Russia stand out from other compared countries in two respects: a) the Czech central state administration appears to be relatively largest in size in terms of percent of population; and b) in proportion to overall number of citizens the Russian public sector employees in health and education appear to outnumber those in other countries.

The extent to which responsibility for job classification is centrally held or decentralized to agency level allows categorizing civil service systems into *rigid position (centralized)* and *flexible position (decentralized).* The Czech civil service can be classified as *flexible position (decentralized)* since until recently job categories in the Czech civil service had not been sharing common attributes and ranking factors.

It appears that the Civil service Act adopted in 2002, if consistently implemented, might widen perspectives for career development of various categories of civil servants. The elements of centralisation have been gradually reinforced since 2001 when the General Directorate of Civil Service within the Office of the Government was established. It has been assigned the task since January 2002 of coordinating human resources management across central state administration (SIGMA, 2003:20). The Government decree No.469//2002 of 5 August 2002

(SIGMA, 2003:12) introduced a Catalogue of Job Positions and Qualification requirements. The decree was enforced starting from 1 January 2004.

Administrative systems can be further distinguished by the extent to which jobs are *broadly or narrowly defined* and the extent of predominance of *generalist or specialist jobs*. As demonstrated earlier in this study, until recently civil servants' jobs in the Czech Republic have been classified into 12 grades as stipulated by the Labour Code. Each grade was divided into 12 steps (classes) according to the length of service. So looking at the situation from this point of view we come to the conclusion that *jobs in the Czech civil service* are *narrowly defined.*

The new civil service law (2002) seems to consolidate this trend as it provides for the existence of 12 grades for civil servants and 16 grades for public employees. The law also distinguishes two large categories of civil servants – *principals* and *subordinates.* The group of *principals* comprises basically positions of state secretaries and their deputies; heads and deputy heads of departments; directors and deputy directors of divisions and sections. Positions of subordinates are divided into two groups: those civil servants who have high education and those with vocational or secondary education (SIGMA, 2003:12).

With regard to prevalence of *generalist* or *specialist* jobs there is a definite *shortage of generalists* in the Czech public administration (NTF, 1998 point 9-4). The Concept of training of Czech civil servants points to a quite insufficient number of professional managers in the Czech public administration who are generalists, i.e. educated and trained in analytical, information, personnel management and similar functions (Concept of Training, 1999, point 4.1).

There are different opinions as to whether the Czech administrative system constitutes a *position based (fixed term) or career (tenure)* employment system. According to Vidlakova the existing employment system is position based rather than career based (Vidlakova, 1999 point 3-1). But she admits that state employees have the right to be promoted according to the length of service in a particular grade and that promotion is automatic according to seniority in a particular grade (ibid, point 3-5). Furthermore, normal employment relationships for the majority of state administration employees were established for an unlimited period of time. Therefore it appears that strong elements of a tenure system have been and still are present in the Czech administrative system: "Tenure is guaranteed to some extent" (SIGMA, 2003:11). Although the new law (2002) does not provide for permanent tenure, the basic type of service contract will be of unlimited duration. According to Onisko the Czech state administration is a "position based system" with elements of career system (Onisko, 2000:7). He argues that advance is guaranteed within each grade (12 steps according to seniority) whereas transfer to another grade can be authorized by the decision of the head of the agency concerned (Onisko, 2000:7).

Therefore, there are good grounds to argue that the Czech civil service system is a *mixed* one, combining elements of position and career based systems. The Civil service Act (2002) appears to strengthen the mixed nature of the existing system although only full enforcement of the law will show how the system functions in practice.

Until recently in the Czech Republic there has been no obligation for state administration agencies and institutions to have comprehensive *job descriptions* for their staff. Only the main activity determining the grade needed to be specified (SIGMA, 2003:11). Therefore proper *modern job description* was a rarity and employees often did not know well the tasks and requirements formally assigned to them (NTF, 1998: 6-2). Following full entry into force of the Civil service Act all recruitment issues, job descriptions and overall classification shall be responsibility of the Director General of the Civil Service.

7.3.2 Job Definition and Classification System - Russian Federation

The total *number of civil servants* of federal state bodies, state bodies of the constituent entities of the Russian Federation, and other state bodies established under the Constitution of the Russian Federation, (hereinafter collectively referred to as the state bodies), together with municipal servants of local self-government bodies amounted to 1,053,100 at the beginning of 2002. This total includes a verified 315,100 civil servants and other employees of federal executive bodies. For the period of 1992-2002, the number of civil servants and other employees of these bodies grew by a factor of 1.8, primarily at the regional level (Federal Programme 'Reform of the Russian Civil Service', 2002). It is almost identical growth to that in the Czech Republic for the same period.

The data provided by the Federal State Statistics Service is somewhat different. In 2002 overall employment by Russian federal and regional government amounted to 1,252,300 persons; in 2004 the figure was 1,318,600 (Federal State Statistics Service, 2005:8). One of the possible reasons for discrepancy in the data is that Federal State Statistics Service counts also technical, support and security staff on payroll of the respective state organs. The years 2005-2006 have seen further increase in the number of state officials up to the level of 1.623 million (Federal State Statistics Service, 2008:4).

The verified number of federal civil servants in central administrations of federal executive bodies amounted to 24,900 and in their territorial bodies to 290,200 (ibid).

The system of civil service personnel management is *decentralised* and continues to be largely based on the rules set up under the Soviet system with a number of new regulations formally introduced but often disregarded. Growing autonomy of regions in the 1990s contributed to weakening of centre-periphery relationships within the civil service. Following adoption in 2006 of the Roster of federal civil service positions, responsibility for job classification at the federal level is now centrally held. Therefore in this respect the system tends to be predominantly of a *rigid* nature.

The new law 'On the State Service System of the Russian Federation' (2003) provides for the establishment of a central organ that should manage the civil service. However, five years after the law was enforced establishment of such a central body was still pending.

The law 'On State Civil Service of the Russian Federation' (2004) stipulates that there should be four categories of civil servants (article 9), five groups of civil service positions, and five civil servants' ranks (article 11) which in their turn are divided into three classes. From this point of view the new law does not differ much from the previous one since only the number and titles of civil servants' categories have been changed. Therefore, we may conclude that jobs in the Russian civil service are rather *narrowly defined.*

The balance between *specialist* and *generalist* positions in the Russian civil service is shifted towards prevalence of the former. However, deficit of professionals in certain areas e.g. analysts, personnel managers is considerable. The proportion of civil servants at the federal and regional levels with basic education in technical, agricultural or natural sciences (that is generally irrelevant to the job needs) was around 30% in 2004 compared to 56% in 1995 (Federal State Statistics Service, 2005:24).

The Law 'On the Basic Principles of the Russian Civil Service' adopted in 1995 specifies that "an individual can join the civil service on the basis of a labour contract to be concluded for an indefinite term or for a term of not more than five years" (Article 21, point 6). Although the authors of the new law (2004) advocated wider use of fixed term contracts neither the text of the law (Article 25) nor subsequent normative acts provide an effective mechanism for practical implementation of this provision. In practice the absolute majority of contracts with 'existing' staff and new entrants to the civil service are of an indefinite length. Advance to the next grade and rank is not directly linked to performance and depends almost entirely on the length of service. Therefore the Russian civil service system belongs to a category of *career (tenure) systems.*

Until recently job descriptions in the Russian civil service were drafted very formally. Preparation of modern *job descriptions* was declared one of the civil service reform priorities. Available information indicates that the overall situation with job descriptions is beginning to change in the positive direction.

7.3.3 Deployment Rules - Czech Republic

In the absence of a specific civil service related legal framework for Czech government employees, *human resource management* remained the domain of individual ministries (World Bank, 1999:5). Hence there exists significant variability in personnel management practices. Personnel management was understood and implemented in practice primarily as personnel record keeping and administration (NTF, 1998: 6-2). Many HR managers saw their role in personnel management only as determining the salaries and bonuses and the approval of reorganization and related transfer of people.

Each ministry and agency performed the role of an employer in relation to its staff members (Onisko, 2000:8). Contracts (either for indefinite term or fixed term) were signed between an individual and the respective agency (Onisko, 2000:7). Thus, civil servants *mobility* was seriously constrained. Further, according to the Labour Code provisions mobility was possible only with the consent of the person concerned (Vidlakova, 1999, point 2-8). Although the new law makes reference to a general principle which implies that civil servants have legal relationship with the state, it also introduces the notion of "sphere of service" for an individual civil servant thus complicating horizontal mobility. Therefore horizontal mobility as a means of professional development and sharing of cross sectoral expertise may be difficult. "Mobility between ministries or from outside the public service is rare and this reduces cross fertilization of ideas and cooperation between ministries and agencies" (SIGMA, 2003:138).

According to Vidlakova "persons with experience gained outside the public service have broad possibilities of *entry to public service*" (Vidlakova 1999, point 2-3). However, there is no reliable information as to the extent such opportunities have been used in practice. According to article 39 of the Civil service Act (2002) transfers are compulsory but for not more than 6 months.

Until recently, general principles of recruitment and promotion of civil servants were regulated by the Labour Code (1965). There was no differentiation in this respect between civil servants and other employees. Explicit rules for promotion did not exist allowing a considerable degree of discretion by managers (SIGMA, 2003:10). NTF report points to "generally disorganized conditions in civil service which do not guarantee professional growth of the employees, career promotion based on defined rules and clear career prospects" (NTF, 1998: 6-2).

The right to promotion has been based mainly on the length of service in the grade (Vidlakova 1999, point 3-1). The promotion practices allowed for a considerable degree of arbitrariness and patronage (SIGMA, 2003:10). Furthermore, considerable discretion in recruitment and promotion, in addition to absence of restrictions to involvement in party politics, undermined principles of impartiality and professional independence.

In the absence of such rules civil servants in their turn were not sufficiently motivated to achieve performance objectives. The new law (2002) envisages that recruitment issues will fall under the prerogative of the Director General of the Civil Service. However, there are no indications that decisions on recruitment and appointment can be subjected to judicial review which is regarded by some experts as an obvious shortcoming of the new law.

Numerous evidence suggests also that promotion into higher grades is often used to counterbalance overall low remuneration granted to civil servants (SIGMA, 2003:11).

Staff turnover in the Czech civil service in 1997 (see table 7.2) varied between 4.75% in the Ministry of Agriculture to 14.7 % in the Ministry of Education, Youth and Sport. (Vidlakova, 1999) but according to other sources this number was larger – from 25 to 30% in 1996 (Onisko, 2000:20).

Table 7.2 Estimated Staff Turnover in Central and Regional Government of Selected Countries (percentage), 2000

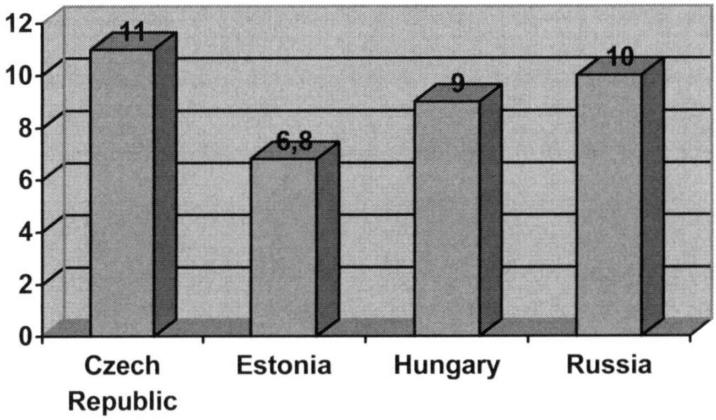

Source: World Bank 2000; Administrative reform, Moscow 2004.

According to the World Bank data *average turnover* below Director Level appears not to be very high (World Bank, 1999:9). In this connection many observers point to *informal rules* which discourage dismissal except for cases of clear misconduct. Ratio of number of staff subordinate to managers averages 5 for central agencies and appears to be rather low (World Bank, 1999:90).

In the Czech Republic funds for *civil servants' training* are not allocated centrally. Each ministry and agency determined how much to spend on training its officials within established limits (Vidlakova, 1999: point 3-7). As a rule fees for training are covered from a special fund for social needs amounting to 2-3% of the total payroll of a particular ministry or agency. However, allocations from this fund may be directed also to purposes other than training. Therefore the real percentage of the total payroll spent on training is considered to be around 1% on average. In general, training was not obligatory for career development (Vidlakova 1999, point 3-9). It is known that the Ministry of Finance spent on average 2% of its wage bill on training (World Bank, 1999:140).

"There is no comprehensive training system that would guarantee a targeted preparation and recruitment of graduates for the civil service" (NTF, 1998: 6-2). A comprehensive system of in-service training of central government officials is also missing (ibid). Responsibility for training of existing staff resides with individual ministries and agencies. "Training of civil servants is not centrally coordinated, it is organized on a case by case basis, responsibility for its effectiveness is not determined" (NTF, 1998: 2-2). Motivation for training at manager level is weak. The number of civil servants who did not meet the educational requirements of their posts in 1999 was around 15% of the total civil service membership (Concept of Training , 1999).

Until recently there has been *no central educational institution* for training and up-grading qualifications of Czech civil servants working in central government. Generally there appears to be a shortage of highly qualified specialists, trainers and lecturers on various aspects of modern public management. Donor and bilateral programmes do not seem to be able to meet in full the demand in this area.

In October 2001 the Government adopted a resolution approving a Strategy of systematic training of civil servants. In September 2001 the Section on Personnel Management and Training was set up in the Office of the Government. An Institute of Public Administration in charge of developing strategies and coordination of training of civil servants was also created in 2002 and organizationally became part of this Section. The establishment of the Institute of Public Administration constitutes a significant milestone in the development of training strategy and practice for the Czech civil servants.

7.3.4 Deployment Rules - Russian Federation

Human resource management in the Russian civil service is largely decentralised. Each ministry or government agency continues to be responsible for hiring and managing its staff. The establishment of a central management organ for the entire civil service, as stipulated in the Law 'On the State Service System of the RF' (adopted in 2003), is still pending. Nonetheless some elements of centralised HR management can be found; for example, in the areas of civil servants training or management of top level positions in the Government and Presidential Administration.

The mechanism of *appointment* to the upper rungs of career ladders is still predominantly very informal. One of its most important elements is the ability of a candidate to establish good personal contacts and to prove one's loyalty to superior management. The general tendency is that the higher the position to be filled is, the less emphasis is placed on the professional qualities of a candidate and his education and the more decisive is the factor of his personal loyalty to the superior. This is particularly true for civil service positions starting from the level of Deputy Head of Department and upwards.

Appointment based on transparent and open competition is still rare. The new legislation (2004) introduced in fact more opportunities for bypassing the requirement for competition as compared to the law from 1995. Although the reform set the objective of competitive based recruitment and promotion, the mechanism for ensuring compliance with these principles is generally weak.

The new law (2004) provides for a possibility of *side entry* to the civil service on the basis of competition. However the procedures and mechanism to ensure implementation of this provision are to be determined by secondary normative acts.

Opportunities *for inter-departmental (lateral) mobility* are generally limited and such cases are rare. Existing practice of deployment of personnel provides for opportunities for movement but mainly within one organisation. The system of rotation of civil servants among federal ministries and between them and other state organs is only now being developed. The practice of temporary assignment of middle and high level officials to regional executive or judicial bodies of power, common in Soviet times, is no longer used.

High turnover of cadres has been one of the serious problems faced by the Russian civil service in the 1990s. Civil service pay at that time was unacceptably low and the prestige of work in the service was falling. During the period 1992-1995 the turnover of employees involved two third of the total staff of executive power agencies at federal level and one third of those at regional level (Ovsyanko, 1996:71). In the period 1991–1994 the staff of the executive branch of federal and regional government in Russia turned over by 58%! (Grishkovets, 2002:53). As a rule it was the best employees who left to the private sector. In 1995, on average 14% of positions were vacant in federal ministries and agencies with vacancies in some of the ministries reaching 40% (Ovsyanko,1996:73).

In the late 1990s the rates of turnover in the Russian civil service slowed down. The problem remained largely at low level positions since advance to senior positions for young personnel was virtually blocked because of utter shortage of vacancies. As a result young people had no stimuli to enter or stay long in the civil service (Gimpelson & Magun, 2004:7). Following the considerable increase in civil servants' pay in recent years turnover of cadres appears to have slowed down even further and remains now at generally acceptable levels.

Attracting new entrants with appropriate skills and values is also improving because of retreat of the older generation.

"Soviet type" traditions are particularly strong in what concerns delegation of authority and rights to lower echelons of civil service. *Decision making* tends to be concentrated in the higher echelons. In the mid 1990s in many ministries the ratio chief - subordinate was approaching 1:1 (Otechestvennyie Zapiski, 2004 :128).

There is no centralised bank of data on the types of skills needed across all levels of government. However, attempts to create such a data bank are inhand.

Retraining and professional development of federal civil servants is currently being funded from the federal budget. Under the Federal Law 'On the Federal Budget of the RF for 2002', 417.5 million roubles was appropriated for this purpose. The sum included 68.5 million roubles allocated for the so called state order that serves as a basis for retraining and professional development of federal civil servants (Federal Programme 'Reform of the Russian Civil Service', 2002:6).

The central *educational institution* for training civil servants is the Russian Academy of Civil Service (RACS) which has a large number of regional branches. In 2004, 15,7% of the total number of civil servants underwent different types of training, of which over 60% were trained by RACS and its centres. Less than 10% of the trainees attended training and upgrading courses which lasted more than one month (Kotchegura, 2006:281).

Standard criteria of eligibility for *training* as well as links between professional up-grading and career prospects generally do not exist. The level of co-ordination of educational institutions' training programmes with personnel departments of federal and regional agencies is generally low. Training of civil servants lacks coherent strategy and until recently has been seriously under-financed in particular at the regional level. The number of civil servants whose professional education does not match the qualification requirements of their positions tends to grow (Federal Programme 'Reform of the Russian Civil Service', 2002:6).

7.3.5 Job Security and Membership Rules - Czech Republic

Pending full enforcement of the Civil service Act, the existing rules and practices do not comprise strict requirements for *recruitment* on the grounds of open competition (SIGMA, 2003:10). Also there was no regulation which required public advertisement of vacancies. Open competition was not obligatory and therefore was regularly used only for separate categories of civil servants or in separate government agencies (SIGMA, 2003:9). For example, the Ministry of Labour and Social Affairs has practiced publication of vacancy announcements in the mass media. Candidates were shortlisted and then interviewed by the commission consisting of 3 to 5 staff members. Recommendations of the commission had to be approved later by the Minister.

According to Onisko, recruitment in most cases lacked previously agreed and approved standard formal procedures (Onisko, 2000:31). The head of a structural unit had total discretion in selecting his subordinates. Probation period was up to three months (Onisko, 2000:6).

There was no possibility for appeal against recruitment and appointment decision except in clear discrimination cases where violation of the Charter of Fundamental Rights and Freedoms was presumed (SIGMA, 2003:10).

In general until recently, recruitment into civil service was characterized by low attractiveness to potential candidates and low transparency of the process.

According to the new Civil service Act (No. 218, 2002) the process of recruitment will incorporate the so called preparation for civil service phase. Selection of candidates for this phase (it lasts 12 months) should be made through public competition. A selection committee is composed of three civil servants and no outside observers are invited. Selection is accomplished mainly by means of an interview.

The probation period ends by oral and written examination conducted by an examination committee consisting of three civil servants and two external experts. The procedure is followed by an appraisal done by the supervisor. As a result of this procedure a candidate is either approved for service or his candidature is rejected.

However, "new recruitment processes are unlikely to be put in place during the transition period, which has now been extended until 31 December 2007" (SIGMA, 2003:5).

The Civil service Act stipulates that *vacancies* across virtually all levels of civil service should be advertised and filled in by means of open competition. This should also contribute to establishment of merit based promotion system. The first round of competition will involve only those who are already employed in the organization offering the vacancy. The second round is held when the first does not identify suitable candidates. It involves both acting civil servants in other organizations in the same sphere of service as well as those persons who have completed the preparation phase for service. Finally, the third round targets civil servants in other organizations beyond the sphere of service concerned. During the transitional period (until full enforcement of the law) current civil servants are supposed to apply for admission to the preparatory civil service in order to be able to keep their positions. The preparation in this case means retraining and passing a service examination.

The SIGMA report 2003 points to a number of flaws in the new law which may in practice undermine principles of promoting professionalism, de-politicisation and reduction of cases of conflict of interest (SIGMA, 2003:15). For instance, the law regulates that a candidate for a civil service position can be rejected if a recruiting manager considers that the individual does not respect democratic principles of the Czech Constitution. In view of many experts this clause may lead to cases of arbitrary and biased decisions (SIGMA, 2003:11).

No restrictions concerning *trade union* activities and the right to strike have existed in the Czech Republic (Vidlakova, 1999, point 3.10). The new Law retains the right of trade union membership and the right to organize strikes with one limitation: civil servants occupying position of Principals.

"All employees appear to work under a cloud of uncertainty regarding the *security* of their posts, which seems to increase after major political changes and to be greater the higher one's position" (World Bank, 1999:81). This is particularly true with regard to top level officials who are often replaced following elections (Onisko, 2000:21).

In general administration has the right to *dismiss* an employee in case of physical incapability, professional incompetence or grave/regular misconduct. Dismissal is also possible following reorganization or abolition of the post. In principle, decisions on dismissal can be appealed in a civil court.

The *pension system* provided to the Czech civil servants does not look sufficiently developed and attractive. It is based on the principle "pay as you go".

7.3.6 Job Security and Membership Rules - Russian Federation

The Russian Constitution (Article 32, point 4) stipulates that "citizens of the Russian Federation have equal access to civil service employment". Apart from general restrictions like Russian citizenship, respective age, good health, and knowledge of the state language any citizen has the right to apply for a civil servant position.

Issues of recruitment to civil service are also regulated by the Law 'On State Civil Service of the RF' approved on 27 July 2004 and Presidential Decree No.112 dated 1 February 2005 'On Competitive Recruitment to State Civil Service of the Russian Federation'. Article 21 of the Law specifies that *"Citizens of the Russian Federation who have reached the age of 18 years, are fluent in the official language of the Russian Federation and meet the relevant qualification requirements established by this Federal Law, shall be entitled to enroll for civil service"*.

Article 22 of the Law sets the requirement of mandatory *competition* in cases of recruitment and promotion to civil service positions whilst stating also a number of exceptions to the rule. Competition may not be held when it concerns:

- appointment to civil service positions with a fixed term in office in the categories "executives" (political appointees) and "counselors (their assistants)";
- appointment to civil service positions in the category "executives" where appointments and dismissals are effected by the President of the Russian Federation or the Government of the Russian Federation;
- conclusion of a fixed-term service contract;
- appointment of a civil servant to another civil service position in the instances envisaged by clause 2 of Article 28 and clauses 1, 2 and 3 of Article 31 hereof;
- appointment to a civil service position of a civil servant (citizen) is included in the personnel reserve formed on a competition basis.

Point 3 of the same article says also that "Competition may be foregone in the case of appointment to certain civil service positions where the performance of job responsibilities involves the use of information that constitutes state secret….".

Regretfully these exceptions to a considerable extent undermine practical application of the principle of competition in the Russian civil service since they allow its bypass all too easily. Particular reference can be made to the clauses dealing with "personnel reserve membership" and "access to state secrets". The point is, that in spite of the law's provisions, *personnel reserve* (a list of civil servants recommended for promotion to high ranking posts) has been so far seldom formed on a competitive basis.

It should be noted that strict requirement to hold a competition for filling a vacant civil service post was also contained in the previous law adopted in 1995 (Article 21) but was rarely observed in practice. Furthermore, if comparison is made between both laws (from 2004 and 1995) the number of exceptions in the new law concerning mandatory competition in cases of recruitment and promotion has actually grown.

The above mentioned Presidential Decree No.112 stipulates that competition for filling a vacant civil service position should be preceded by *publication of announcement* of the vacancy in no less than one periodically printed publication as well as on the web-site of the government structure in question.

Whilst the new Law provides for broader opportunities of *side entry* to civil service and wider application of *fixed term contracts,* the mechanism for the practical application has not been developed yet.

Although the situation with regard to application of competition procedures in the Russian civil service is gradually improving, until recently enrolment and promotion to a higher position in the majority of cases was taking place without competition. According to a representative poll carried out by Gimpelson & Magun in 2002, only 4% of the interviewed civil servants (age up to 35) both at the federal and regional level had progressed through *competition* or special tests at the time of enrollment (Gimpelson & Magun, 2004:11). The poll confirmed that one of the main criteria for recruitment continued to be personal relationship and loyalty to the management. The poll results show that professional qualities and belonging to a patronage sealed network of like minded officials have approximately the same impact on the prospects of career advance (Gimpelson & Magun, 2004:21).

Contrary to the Czech legislation the Russian Law does not grant civil servants the *right for strike*.

Significant politicisation of the Russian civil service complemented by frequent reorganisations of government institutions place stability and security of staff into dependence on political decisions. This is especially true for the civil service staff at the regional level. During the period 1991–2000 the structure of the federal government in Russia had been significantly changed nine times. Under such conditions it is difficult to speak about any stability (Grishkovets, 2002:73).

Any state institution can effectively *dismiss* a civil servant in line with numerous provisions contained in the law. Regretfully dismissals are not done always in strict conformity with the law especially during large scale reorganisations and even more so during reshuffles of political leadership.

However a civil servant has the right to appeal to a court against such decision if he/she is convinced that there is unfair treatment. In practice cases when the court rules in favour of a dismissed civil servant are far from being exceptions.

The existing *pension system* for Russian civil servants possesses certain comparative advantages to the schemes of most other large social groups.

The absolute majority of contracts with 'existing' staff and new entrants to the civil service are of a non-specified duration. Therefore the Russian civil service system can be classified as that of a career *(tenured) system*.

7.3.7 Reward Mechanism - Czech Republic

Until recently, the remuneration system for Czech civil servants (Law 143/1992) comprised a considerable element of bonuses, premiums and extra payments additional to basic salary. The premium can amount up to 100% of take home pay for grades 9-12, and up to 40% for grades 1-8 (SIGMA, 2003:15). In practice, on average 70% of civil servants' remuneration is salary and 30% is additions and premium (Onisko, 2000:8).

The average gross monthly earning in the state administration was 11,788 crowns in 1999 equivalent to approximately 370 USD (World Bank, 1999:88). In 2001–2002 the *average salary* in state administration amounted to 17,000–18,000 CZK (approx. 550 euros). The lowest and the highest salaries were in the range of 9,000 CZK and 37,000 CZK respectively (SIGMA, 2003:16).

Salary and most additions are permanent depending on the position, length of service etc. Various supplements (premium) are granted on the basis of discretional assessment done usually by the Head of unit (service, agency). The criteria which are used to define the size of premium are neither sufficiently specified nor are open for external scrutiny, nor are unified across various government institutions.

Table 7.3 Percentage of Civil Servants' Earnings in the Form of Allowances/Bonuses (2000)

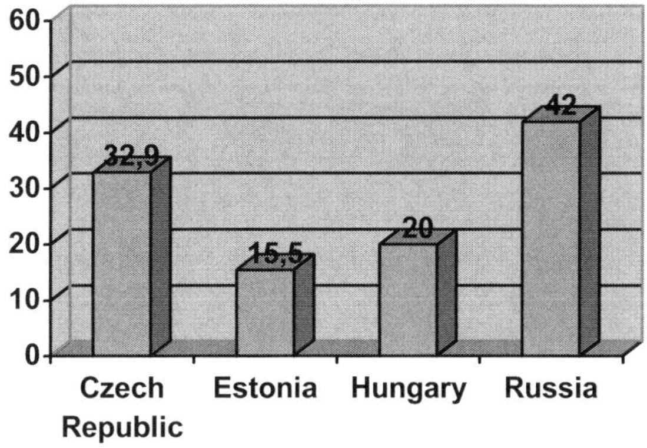

Sources: Ministries of Labor (Czech Republic, Hungary, Estonia); Administrative reform, Moscow, 2004.

Until recently jobs were remunerated differently in the various state institutions. This may change if a Civil service Act is fully enforced. In accordance with Government order 48/1995 the opportunities to use savings from unfilled vacancies as monetary incentive for existing staff were restricted. However, in 1997 these limitations were removed and the savings are more frequently used for general salary increases.

In 1997 budget expenditure to cover wages of employed in the budgetary organizations amounted to roughly 100 billion CZK including allocations to funds for social and cultural needs (NTF, 1998, 5-3).

Table 7.4 Central Government Wage Bill as Percent of GDP

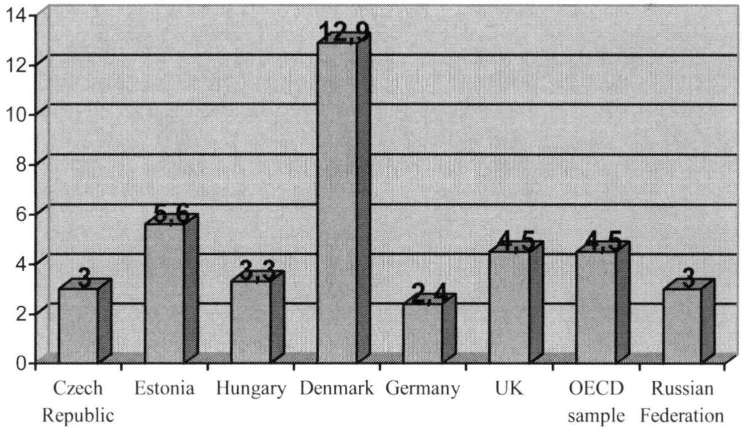

Source: An International Statistical Survey of Government Employment and Wages, World Bank, 1997. IMF (Czech Republic), World Bank (Estonia), World Bank (Russia).

There is no general regulation prescribing *performance appraisal* and its manner (Vidlakova 1999, point 3-6). In general, standard and transparent criteria for *performance based remuneration* are lacking.

In comparison with similar jobs in the private sector, monthly salaries in the Czech civil service are from 25 to 30 % for director level positions but increase up to 60-80 % for lower skilled positions (World Bank, 1999:71).

Salary differential *(compression)* within the system of 12 grades ranges from one to five (SIGMA, 2003:16). Similarly a World Bank report shows that compression ratio of highest to lowest salaries in the Czech state administration is between 4.4 and 5.1 (World Bank, 1999:71). However, taking into account highly discretionary allowances and supplements reaching up to 100% for the Czech top level servants, this ratio may reach 7-8. (World Bank, 1999:71). In comparison the average compression range in EU member states is from 6 to 10.

Table 7.5. Compression Ratios (basic civil servants' wages).

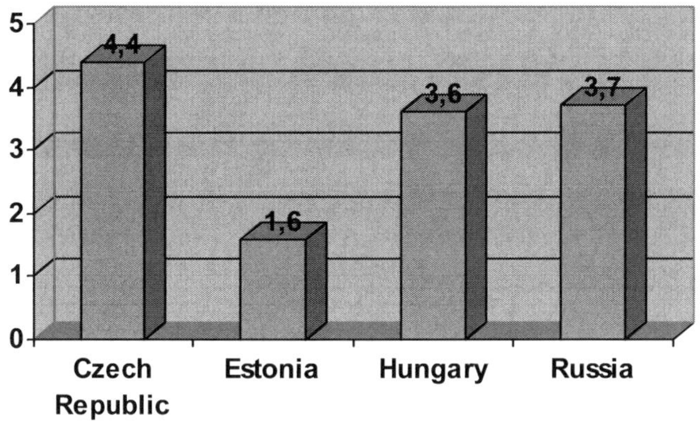

Source: Official government wage scales, calculations of World Bank staff (Moscow); Administrative reform in Russia, 2004.

The new Law introduces a unified salary system and a unified classification of civil servants (National Catalogue of Job Positions) which however will be fully applied only after the transition period is over. The new system envisages that remuneration of a civil servant will be based on tariff classes (linked to pay grades and pay classes) and regular performance appraisal. It is expected that its introduction will result in overall 40% increase in salary levels across the service. Some experts reckon that the Law grants to Czech civil servants "disproportionate privileges in the form of generous fringe benefits and perquisites" (SIGMA, 2003:5).

The salary will consist of two basic components: entitlements and discretionary allowances. Apart from basic salary the first component comprises wherever appropriate extra pay for service, management duties, overtime, work in difficult conditions and some other criteria. The second component is composed of two allowances – personal extra pay and reward for extraordinary performance. The first allowance can amount to 20% of basic salary and the second allowance up to 100% of basic salary. The future system incorporates an extensive list of supplements and allowances which, in the absence of clearly defined rules for their application, may strengthen discretionary powers of top managers. This increases the risk of arbitrary decisions and politicization (SIGMA, 2003:18).

In general the new wage system is aimed at widening salary differentials among jobs in Czech state administration and better rewarding selected highly qualified specialists (Dvorakova, 2005:3).

7.3.8 Reward Mechanism - Russian Federation

In the 1990s the compensation system of civil servants in Russia was based on the provisions of the Federal Law No. 119 of 31 July, 1995 "On the Basic Principles of the Russian Civil Service". Article 17 of this Law stipulated that civil servants' pay shall comprise a base salary and increments in accordance with the assigned civil service class rank, special conditions of civil service, length of civil service and also bonuses based on job performance. In practice this system of civil servants' remuneration gave definite priority to the principles of seniority and length of service.

Prior to the default in 1998 the *average civil servant's monthly salary* in Russia was about $450 but the default reduced it by over 300 per cent. In the year 2001 a civil servant's average monthly salary in Russia amounted to 4,800 roubles (approx. $180) (Federal programme "Reform of the Civil Service of the Russian Federation", 2002). Various types of bonuses and allowances constituted from 35% of the total amount of compensation for mid and low level civil servants and up to 200% for high level officials.

One of the declared priorities of the programme of reform of the Russian civil service (adopted in 2002) was introduction of a *performance based pay*. By 2007 these plans had resulted in a number of experiments in selected ministries and regions. As in the past, the level of compensation of the majority of civil servants does not depend directly on the results achieved. There exists substantial differentiation in the level of civil servants' compensation across the various regions of Russia.

In the years 2004–2006 the average pay (remuneration) of civil servants in the federal government structures has grown over 100%. In 2006 average monthly remuneration of federal civil service employees was estimated to be 13,425 roubles (approx. USD 500). This is higher than general Russia monthly average of 10,728 roubles ("Kommersant Daily", 27 March 2007). Average monthly pay of a Moscow based federal civil servant was 24,124 roubles (approx. 800 USD) in 2006 (ibid).

The pay of top level civil servants (heads and deputy heads of departments, deputy ministers) has been raised from 300% to 500% for the same period. It is noteworthy that the base-salary, transparent to society and the tax service, was not increased significantly in comparison with various types of bonuses (Nezavisimaya Gazeta, 6 February 2006).

Federal Law No. 79 "On the State Civil Service in the Russian Federation" adopted in the year 2004 and in force as of 1 February 2005 stipulates that "The pay of a civil servant shall include the monthly civil servant's salary in accordance with the occupied civil service position and the monthly civil servant's allowance in accordance with the assigned civil service class ranking, which shall jointly make up the monthly pay of a civil servant, and monthly or other additional payments" (Article 50 of the Law).

The same article refers to six types of additional payments covering: length of service (monthly increment to base salary up to 30%); special work conditions (monthly increment to base salary up to 200%); working with state secret data (as established by the legislation of the RF); bonuses for completing especially important and difficult tasks (no maximum limit shall apply); monthly bonuses; lump-sum payment upon granting of annual paid vacation; and financial assistance paid from the civil servants payroll fund.

The Law also sets out some novelties. In particular the law rules that for specific civil service positions, special compensation terms may be established, with such compensation paid depending on professional service performance and efficiency indicators specified in the fixed-term service contract.

Compression ratio in the Russian civil service is estimated by various sources to have been around 1/3-1/4.

The new Law has left discretionary allowances virtually undisturbed and they keep on occupying a significant part of a civil servant's take home pay. This undermines considerably the spirit of reform and will continue to feed up patronage and politicisation of the Russian civil service.

According to the official statistics state administration related expenses amounted to 2.8% of the overall expenditure part of the federal budget in 2004 (Federal State Statistics Service, 2005).

7.4 Representativeness

Representativeness appears to be a term which is associated with a wide range of ideas, views and policies. To a considerable extent this may be explained by marked differences in political, societal and administrative values typical of various country specific settings (Meer & Roborgh, 1996:119). These values have a profound impact on the development and function of civil service systems in respective societies.

The study of representativeness in a particular environment helps to understand some important elements of the system of governance and specifically relationships between "the rulers and the ruled". "Representativeness must be viewed against the background of basic concepts of the *raison d'etre* of the state and more specifically the changing relationship between the population and the state" (Meer & Roborgh, 1996:121).

Proceeding from the above reflections we will consider the status of representativeness in the Czech Republic and the Russian Federation above all from the point of view of :

a) demographic (gender, religion, language, ethnicity, educational background) composition;
b) equal opportunity (the degree the recruitment process to the civil service is impartial and based upon functional and standard criteria);

c) interest representation (extent to which particular groups of civil servants act as sponsors of interest groups);
d) responsiveness and accountability to external actors as important elements of representativeness.

The analysis of these elements and characteristics of representativeness provides insight into the extent of legitimacy of civil service and control of civil servants by society. The idea is that acceptance and control of administrative activities will increase when the composition of civil service more or less corresponds to that of society in terms of its demographic composition, opinions and interests (Meer & Roborgh, 1996:123). In essence representativeness may serve as a factor containing possible abuse of power by civil servants.

Extent of responsiveness and accountability are crucial for grasping the relationship between citizens and civil service. It is also essential to find out to whom accountability and responsiveness are exercised: to society, to a particular societal group, to a party, to government or to president.

7.4.1 Demographic Representation - Czech Republic

Available and scarce data lead to the conclusion that there is no large discrepancy between the demographic composition of the Czech civil service and the rest of the society as far as nationality, religious affiliation and language are concerned. For obvious reasons the civil service personnel is more educated than many other social groups in the society. Expert opinions indicate also that representation of the younger generation in the Czech civil service is lower than its part in the composition of the Czech entire population. Problems in recruiting young staff stem generally from the rather low prestige of the civil service and numerous opportunities appearing in the more lucrative private sector.

The ages pyramid in the Czech civil service shows a deficit of the officials aged 30-45 years (SIGMA, 2003:8). This means that that the Czech civil service will face a serious problem of finding a skilled replacement for a considerable amount of experienced cadres who will retire in the coming 5-10 years.

In the Czech public administration women considerably outnumber men: 300,000 against 73,000 persons respectively (Czech Statistical Office, 2005). However, at the top level positions women are seriously underrepresented. Average age of women occupying positions of managers in the public and defence sector is 42.6 years whilst the age of men in comparable positions is 46.3 years (Krause, 2005:30).

7.4.2 Demographic Representation - Russian Federation

In 2002 the average age of civil servants in the Russian civil service was 40 years (Federal programme "Reform of the Civil Service of the Russian Federation", 2002:5). The existing gender and age structure of civil service personnel has a number of peculiarities which distinguish it from the gender and age structure of Russia's employed population.

In the same year 2002 women accounted for over 70 percent of all civil servants including 71.8 percent in executive bodies, 67.6 percent in administrations of judicial bodies and prosecutors' offices, 55.8 percent in administrations of legislative bodies, and 55.4 percent in administrations of other state bodies. The high proportion of women in the civil service is explained by their prevalence in low and mid level positions in state administration (Federal programme "Reform of the Civil Service of the Russian Federation, 2002:5). Women are however seriously underrepresented at higher levels of civil service.

Throughout the 1990s there was substantial deficit of young specialists in the civil service which persists at present though to a lesser degree. In 2002 more than 65% cadres in the federal civil service were over 40 years of age. It is estimated that within the next 10 years over half of civil servants employed in key positions will retire and it will be hardly possible to fill these positions with specialists possessing the necessary experience and professional skills (Otechestvennyie Zapiski, 2004 :153).

7.4.3 Equal Opportunity - Czech Republic

The Civil service Act (No. 218, 2002) includes provisions that ensure equal treatment of civil servants and candidates for civil service positions. In particular, it stipulates that:

(1) A service authority shall be obliged to ensure equal treatment of all public servants as regards the conditions of their performance of service, remuneration and other monetary fulfilment, education and opportunities to achieve promotion in service, unless laid down otherwise by the law (Part 6, Chapter 1, article 80).

(2) "Any discrimination of public servants whatsoever in service relations pursuant to this Act for the reason of race, colour of skin, sex, sexual orientation, language, belief and religion, political or other opinion, membership or activities in political parties or political movements, trade unions and other associations, nationality, ethnical or social origin, property, family, state of health, age, family status or obligations towards the family shall be prohibited. Conduct that is of a discriminatory nature, although not directly, but rather in its consequences, shall also be prohibited" (ibid).

There is no information available as to possible serious violations of equal opportunity rights in the Czech Republic with regard to civil service employment except concern voiced by a number of non-government organisations about highly politicized decisions on appointment to mid and top level civil service positions.

7.4.4 Equal Opportunity - Russian Federation

The Federal Law "On the Basic Principles of the Russian Civil Service" adopted on 5 July 1995 provided for equal treatment of candidates for civil service positions. Article 21, point 2 of the law stipulated that "It is not allowed to establish any direct or indirect limitations or privileges for those entering the civil service depending on the sex, race, nationality, language, origin, property and official status, place of residence, lack of availability of citizenship of the subjects of the Russian Federation, religious affiliation, convictions, membership in public associations….."

Similarly the new law "On the State Civil Service of the Russian Federation" adopted on 27 July 2004 states in Article 4 point 21 that "Citizens of the Russian Federation who have reached the age of 18 years, are fluent in the official language of the Russian Federation and meet the relevant qualification requirements established by this Federal Law, shall be entitled to enroll for civil service".

In practice there are known cases of violation of the existing legislation where regional authorities have introduced additional criteria for those seeking jobs in the civil service. For example, candidates for civil service positions in the Republic of Tatarstan are required to demonstrate knowledge of the Tatar language.

7.4.5 Interest Representation - Czech Republic

Analysis of opinions and information from various sources provides sufficient ground for claims that many Czech government officials still think that citizens exist for them, not the other way round (NTF, 1998: 6-2). The image of civil servants in the Czech Republic is generally not favourable. According to the survey conducted in 2007 by an independent research agency most frequent association evoked by an average Czech state official among its compatriots is that of "an unpleasant arrogant person" (Westminster, 2007).

In the absence of clearly articulated policy objectives, mission statements and sound legal framework Czech civil servants have been often either unaware of the possible "conflict of interest" or under no bounding pressure to observe its principles. "The policy missions of the public administration have not been clearly defined to incorporate citizens' needs and rights, and competition and market openness principles" (OECD, 2001 : 136).

There is enough evidence to conclude that "defence of corporate and vested interests" is by no means an exceptional and rare phenomena particularly among Czech high ranking civil servants. "It is quite usual that people leave positions in high politics or state administration directly to top level positions in big companies or consultancy firms – often those that belonged to the field of their competence" (Konopasek, 2002:21).

Even so, examples of institutions which are regarded as exemplary in defending citizens' rights and interests do exist in the Czech Republic. According to official statistics during the period 2000–2004 the institution of Ombudsman has dealt with over 14,000 citizens' complaints. Of these there was resolution in half the cases and the remaining 7000 were provided with legal advice. Opinion polls conducted in 2004 showed that Mr. Motejl, the Czech Ombudsman, was the second most trusted public figure in the country. With a 53% approval rating he was second only to the President and considered more trustworthy than any cabinet minister including the Prime minister (Radio Prague 14 October 2004).

7.4.6 Interest Representation - Russian Federation

Since the times of the tsars and emperors the Russian bureaucracy has been known for its readiness to serve the powerful and disregard the needs and interests of ordinary citizens. These traditions have largely survived till present in spite of fundamental changes in the political and economic system of the country.

Successful career advance of a Russian civil servant often depends on the external support provided by a "sponsor" - usually a commercial entity or a "patron" which could be an influential politician or high ranking official (Afanasyev, 1998). Whose interests are promoted and defended by these civil servants is clear enough.

The whole post-communist experience of Russia shows that it is quite difficult to move away from the Soviet concept of the person serving the state (in essence serving the political elite) to the democratic principle of the state serving the person.

In addition, civil service practices in recent years have provided numerous examples when the corporate interest of the bureaucracy prevailed over the interests of the state and society.

Finally, in the atmosphere of endemic corruption many civil servants have actually "commercialized" their civil service posts. The phenomenon of extracting the so called "status rent' by state officials out of their 'special' relations with businesses is widely spread and has been criticized by President Putin on numerous occasions. In goes without saying that in such a situation the principal objective and interest of a civil servant remains the size of the "profit" he/she can acquire.

Within the framework of current civil service reform, measures to strengthen impartial execution of the policy by civil servants and eliminate possibilities for arbitrary use of public authority have been neither consistent nor effective.

7.4.7 Accountability - Czech Republic

As we have already pointed out in Chapter V there does not exist a precise term for the word 'accountability' in the Czech language. "Public accountability does not have a simple and well-established equivalent in the Czech language" (Konopasek, 2002:7).

The findings of one of the projects financed by the European Commission in the Czech Republic acknowledge that "Responsibility (accountability) of state officials exists vertically, but not horizontally. From the point of view of citizens it does not exist at all" (Phare, 2000:7). Similarly, a World Bank publication states that "institutional arrangements to hold public servants and government agencies accountable are weak in the Czech Republic" (World Bank 1999:71).

According to a report on public accountability in the Czech Republic at the beginning of the post-communist transformation "political leaders enjoyed unquestioned legitimacy and emphasized economic aspects of the reform. The reforms of regulatory frameworks and of control mechanisms were delayed. The delay, together with a lack of competence and easy access to huge financial resources, led to proliferation of corruption, economic crime and loss of trust of the public as well as of foreign entrepreneurs. Despite current effort, results of the fight against frauds and corruption have not been convincing" (Konopasek, 2002:3). The same author argues that weak accountability of Czech politicians and public authorities resulted in such phenomena as uncontrolled links between politics and business, power monopoly of the two most influential political parties based on so called Opposition Treaty, the split of Czechoslovakia in 1993, non-transparent privatization, lagging and protracted reform of civil service, corruption in public administration, and inefficient law enforcement.

The Administrative Procedure Act (1967) protects to a certain extent citizens' rights of appeal and specifies duties to citizens of public administration representatives in such cases. However the Act does not apply to all administrative decisions which have impact on individuals. The accountability mechanisms of the bureaucracy envisaged by the Act are weak and appeals procedures are costly (OECD, 2001:52).

7.4.8 Accountability - Russian Federation

In spite of profound changes in the country's political system since the late 1980s effective control over state administration by the society is largely absent. To a considerable extent the bureaucracy does not bear external responsibility for its actions or lack of actions and hence is under no real pressure to improve performance and responsibility. In the opinion of M. Dmitriev, Russia's former first deputy minister for economic development, "the administration was never accountable to the general public" ("International Herald Tribune", 19 July 2006). In the post-communist Russia the state administration was freed from the former strict control by the party and in the absence of effective accountability distanced itself further from society (Obolonsky, 1999:308).

As before, "decisions in the executive branch in essence are taken collectively, nobody carries personal responsibility" (Otechestvenyie Zapiski, 2004:86). Systematic arrangements to hold public servants accountable for their performance are largely absent (World Bank, 2002:16).

Furthermore, in the atmosphere of entrenched patronage, responsibility and accountability of civil servants to society and citizens has been replaced by accountability to those who granted them their civil service posts. Only 15 percent of the Russian civil servants consider control by civil society as effective means of improving civil service activities (RACS, 2003:16).

Introduction of citizens' right to a pre-judicial appeal against an administrative ruling has been one of the administrative reform objectives. However, the process of elaboration and approval of respective legislation has been very slow. There are concerns that this legislation may repeat the fate of the draft law "On Access to Information about Government Activities" which was prepared in 2002 and has been circulating since then along the corridors of power without notable advance to the final approval (Dmitriev, 2004).

7.5 Politicization

This section deals with the extent of demarcation, interdependence and influence between politics and administration. The processes of politicization of bureaucracy and bureaucratization of politics are known for having exerted a notable influence on the development and functioning of civil services in various countries. "Politics shapes administration and administration shapes politics" (Toonen, 2001:183).

Politicization usually refers to levels and types of political activity undertaken by civil servants and attempts by others to politically influence the behavior of civil servants (Aberbach and Rockman, 1987). It is generally acknowledged that the impact of politicization of the civil services has been particularly strong in the post-communist countries. Below we will consider the situation in this respect in the Russian Federation and the Czech Republic.

7.5.1 Politicization - Czech Republic

In the Czech Republic legally defined distinction between political appointments and career civil servants has been absent for quite a long time. "This absence (of legal distinction between political appointments and civil servants – A.K.) undermines professionalism of the staff and encourages civil servants to be politically focused and discourages them from being responsive to cross-cutting and medium term objectives" (Jalovecká, 2005:9).

The Czech post-communist experience shows that political parties that won elections offered public service positions to their supporters and provided benefits to these collaborators in the form of access to various government posts and services. More importantly, often the new governments acted as clients of the patrons who helped the governing parties win elections.

High level of politicization of the Czech civil service to a considerable extent determines actual conduct, attitudes and values of many civil servants who "toss about in conflict between their careers and their pretended loyalty to political parties in hope of getting lifelong privileges for their merits" (Potucek, 2002:48).

A SIGMA/OECD report on the Czech Republic emphasizes "the system allows politicians to create political clienteles within the administration in an unrestricted way, as there are no legally established limits for political involvement of civil servants" (SIGMA, 2003:10).

In the opinion of Danish experts L Moller and O Norgaard, who studied reform capacity of the Czech and Hungarian telecom administrations, many problems in the development of the Czech telecom industry could be explained by the fact that, contrary to Hungary, most important positions in the management of the Czech telecom industry were filled in by politically appointed officials (Moller and Norgaard, 2002:17).

Overall "the unsatisfactory status of the Czech state administration reflects negative implications of long-term political interference in its affairs" (Onisko, 2000:36).

7.5.2 Politicisation - Russian Federation

The values, norms and traditions of governance and public administration existing in Russia favour the creation of a civil service which remains an obedient tool in the hands of the political elite. Total subordination of the state administration to the political authority is deeply rooted in Russia and dates back to the period of the establishment of unified Russian state. A former high ranking Russian official observed "Since the formation of a centralized state about 800 years ago, bureaucrats have mainly served the country's leaders, not its people. And they have been anything but civil." ("International Herald Tribune" 19 July 2006).

Differentiation into political and career tenure posts in the Russian civil service has not been always clear-cut. As a rule, top level civil servants find themselves very close to the policy making process and establish partisan links with the ruling political elite. Almost common practice is involvement of civil servants in the elections campaigns of governors, mayors, heads of administration, and deputies to the Parliament. For instance, a number of federal agencies were providing active indirect support to the pro-government party "Our Home Russia" during 1995 Parliamentary election campaign. Moreover, the principle of "non-partisanship" was dropped in the new Law "On the State Civil Service of the Russian Federation" adopted in 2004.

Changes of political leaders, government, party or coalition in power as a rule have a direct impact on the civil service cadre. "The political nature of administration and dependency on individual directors make the position of civil servants unstable. The weakness of a merit-based system and staff instability partially explain the unprofessional features in administrative operations and unpredictability of the results. These features particularly derive from the fact that career officials are also often replaced on political grounds." (Larjavaara, 2001:16).

On the other hand the bureaucracy in Russia retains a considerable policy making capacity and continues to be a major player in selecting and supporting candidates for key political posts. In Russia the strength and influence of a politician are determined to a considerable extent by the number of high-ranking supporters he/she has in key administrative positions.

7.6 Public Opinion

This section comprises an overview of the dominant perception of the Russian and Czech civil services and civil servants prevailing among the public and citizens. Accent has been made on the general (based on the prevailing society opinions and stereotypes) and individual (based on personal encounters) perceptions of civil servants.

7.6.1 Public Opinion - Czech Republic

At present the attitude of Czech citizens to state and government institutions can be characterized as a certain crisis of trust. The social distance between citizens and the state administration appears to have grown in recent years and in some cases has become critical.

Throughout the 1990s the prestige of the Czech civil service in the Czech society was rather low. The dominant public opinion was that the number of civil servants was excessive, they were ineffective and too many of them took bribes. An opinion poll carried out in 2000 in the Czech Republic by Transparency International, an organisation dealing with corruption, showed that one third of those polled thought that without bribes the Czech state administration would cease to function. Twenty four percent of respondents admitted that in the past three years they had been in a situation where a bribe was expected. (Radio Prague, 14 February 2001).

There does not exist big difference in perception of civil servants at general and individual levels. In a survey conducted by the IVVM (non government research institute) in March 1998, 48% of interviewees stated that they had met with an unjust decision of the authorities. Negative experience was expressed primarily by people of productive age, with higher education and by entrepreneurs (Public Management Forum, 1999).

7.6.2 Public Opinion - Russian Federation

In Russia the image of civil servants has been shaped for years by such wide spread phenomena as corruption and their arbitrariness towards citizens. In the 1990s the distance between civil society and the state administration is reported to have grown further. In mass consciousness civil servant continues to be associated with "the powers that be" and is counter-opposed to common people (RACS, 2005:36).

The survey conducted in 2004 by the Foundation "Public Opinion" among citizens of various regions in Russia revealed that 71% of respondents assessed civil servants' activities highly negatively (Nezavisimaya Gazeta, 2 June 2006). Only 15% of the interviewed held the view that civil servants now work better than 15-20 years ago and almost 50% thought that they worked definitely worse. However, 70% agreed that to be a civil servant was prestigious (a stark change compared to the situation 8-10 years ago).

Results of the poll conducted by the Institute of Sociology of the Russian Academy of Science in 2005 show that half of the population (49,9%) share opinion that corruption among the national economic and political elite constitutes the main obstacle to rapid economic advance of the country (Russian Academy of Sciences, 2005:8).

Findings of another poll revealed that only 18% of respondents (representing various groups of the population) could explain what civil service reform was about, 40% had heard something and 39% learnt about it for the first time (RACS, 2005:37).

7.7 Conclusions

The findings of our analysis conducted in line with the selected methodology for comparative studies (Protocol for comparative studies of national civil service systems) enable making some significant conclusions concerning the studied aspects of the civil service systems in the Russian Federation and the Czech Republic. They also allow identify commonalities and distinctions in the key characteristics of the two civil service systems.

The two civil service systems have certain *common features* which distinguish them from modern administrative systems in the majority of EC member states. In comparison to many states with developed institutional systems we have established that the Russian and Czech civil service systems possess the following peculiarities in common:

- modern principles of performance based remuneration of civil servants are largely absent;
- the part of civil servants' earnings comprising allowances and supplements compared to basic salary is considerably higher than respective ratio in civil services of most industrially developed countries;
- the heads of units have large discretionary power in determining take home pay of their subordinates;
- appointment to civil service positions on a competitive basis has been an exception rather than a rule (often competition is formal with the decision already known before it starts);
- modern and detailed job descriptions have been almost entirely absent until recent reform efforts;
- the level of corruption is much higher in comparison with both EC and CEE average;
- civil servants salaries have been absolutely not competitive with salaries of respective positions in the private sector;

- the start of most institutional reforms in Russia and the Czech Republic has been delayed as compared to the majority of advanced CEE countries;
- training of civil servants has been underfinanced and exposure to best practice has been limited;
- promotion and recruitment to high level posts in the Czech and Russian civil services are highly politicized; there are practically no restrictions to penetration of party politics into the activities of civil service; this undermines principles of impartiality and professional independence of civil servants;
- there is crisis of trust of citizens in relation to central state administration and government;
- there has been a remarkable similarity in the growth of overall number of public administration employees in both Russia and the Czech Republic in the 1990s - roughly 1.8 times over a decade (1993 – 2003);
- the quality and level of professionalism of civil service in both Russia and the Czech Republic have dropped considerably since the late 1980s;
- performance orientation and accountability to the society and various stakeholder groups are underdeveloped in both civil services;
- enforcement of civil service legislation is generally weak and ineffective;
- there has not existed sufficient integrity and uniformity of both civil service systems (e.g. until recently no single employer countrywide).

We have identified also definite variations between the two civil service systems:

- from the legal point of view there was no civil service in the Czech Republic until 2002 whereas in Russia it exists since 1993;
- there was no central training institution for civil servants in the Czech Republic (except for municipal servants) until 2002 whereas the Russian Academy of Civil Service was established in 1995;
- although far from exemplary the effectiveness of the Czech government's activities has been overall higher than that of the Russian government;
- the Czech civil service system has started moving towards a mixed model with elements of both position based and tenure system (Civil service Act 2002), whereas the Russian system is much more oriented towards incorporating strong elements of career (tenure) system, though some innovations, like introduction of short term contracts, are envisaged by the new Law (2004);
- re-training of civil servants has been generally more regular and better organized in Russia compared to the Czech Republic;
- although in terms of overall number of the public sector and civil service employees the Czech Republic and Russia do not stand out much from other industrialized nations they are to a considerable extent different in two respects: a) the Czech central state administration appears to be relatively (per cent to the population) largest in size; and b) Russian public sector employees in health and education sectors appear to outnumber respective employees in other countries as per cent to the population;
- corruption in the Russian state administration appears to have adopted a greater scale than in the Czech state administration, though it continues to be a serious problem in the latter.

These are major observations concerning key features typical of the Russian and the Czech civil service systems. Overall there appears to be more similarities than differences in the Russian and Czech civil service systems as far as their basic characteristics are concerned. Neither Russian nor Czech civil service fits well either of the two types of civil service systems – classical Weberian or modern professional and responsible civil service.

Our research has identified clear deficit of meritocratic principles in the organization and functioning of the two civil service systems. Lack of proper attention to the enforcement of these principles encourages political allegiance and patronage and at the same time undermines professionalism, impartiality and external accountability of the Russian and the Czech civil services.

CHAPTER VIII
COMPARISON OF CIVIL SERVICE REFORMS IN RUSSIA AND THE CZECH REPUBLIC: DYNAMICS, DIMENSIONS AND EXPLANATORY FACTORS

8.1 Introduction

Our methodological approach presupposes that the process of institutional change is subject to the influence of numerous endogenous and exogenous factors. In line with this approach we intend in this chapter to highlight those factors which have had a distinct, often decisive, impact on the initiation and implementation of civil service reforms in each of the selected countries.

The study of civil service systems and reforms in the Russian Federation and the Czech Republic carried out in the previous chapters allows define factors which have performed as either intervening or determining variables. The identification and analysis of these factors is important for understanding the dynamics of the reform, the encountered difficulties and results achieved so far. It is also crucial for the appropriate comparison of the two reforms and exploration of the causes of possible commonalities and differences in reform implementation between the two countries.

The above mentioned factors have been divided into two large groups - *intra bureaucracy* and *extra bureaucracy* factors. This division is entirely conventional and serves mainly the purpose of logical presentation of the research inferences.

First, we look at the factors "arising from within the system" or "linked to the logic of reform preparation and development". These include clientelistic and patronage networks, corruption, bureaucratic resistance, administrative traditions and culture, mislaid reform priorities and inadequate preparation of reform. Secondly, we examine a range of "external influences" such as constitutional and political framework, economic environment, historical legacy, politicization of civil service, budgetary constraints, political leadership and civil society, and impact of "imported" ideas and concepts.

The second part of the chapter deals with the comparison of civil service reforms in Russia and the Czech Republic. Particular emphasis is made on matching capacities for reform, reforms' dimensions and dynamics, driving forces, implementation constraints and results achieved so far. Similarities and distinctions across the range of raised issues are separated out and preliminary findings as to the impact of selected variables on the processes of reform preparation and implementation are developed and summarized.

8.2 Intra Bureaucracy Factors

8.2.1 Resistance of the Bureaucracy

Taking into account some principal features of the bureaucracy in general and of state bureaucracies in Russia and the Czech Republic in particular (as described in the previous chapters), it is unlikely that most civil servants in these countries will support reforms meant to expose them to public scrutiny, enhance accountability, reduce discretionary power and rent seeking opportunities.

It is not easy to substantiate any judgment about the extent of bureaucratic resistance by empirical data since such resistance is always either hidden or masked. Therefore, conclusions are usually based on close observation of behavior, indirect evidence, questioning of experts and thoroughly prepared interviews of civil servants.

One of the useful instruments in this respect is the measure the civil service reform is accepted or approved by civil servants and this can be identified through various surveys.

Such a survey was conducted in November 2003 by the Russian Academy of Civil Service (RACS). It shows that only 11% of interviewed civil servants were ready to fully support administrative and civil service reforms, 79% were ready to do it only partially and 8.5% were not supportive of the reform. (RACS, 2003:17).

Another indicator is the reluctance of state administrators to accept improvements and measures outlined in the proposed reform concept and programme. Quite demonstrative is the extent some radical innovations contained in the draft Russian law 'On State Civil Service of the RF' (the law was finally adopted in July 2004) have been cancelled, distorted or ignored at different stages of the approval and implementation of this legal act. For example, the mandatory requirement to hold competition, when a civil servant vacancy needs to be filled in, contained in the first draft of the Law, was complemented by numerous exceptions in the final document. Another well known case concerns a draft law "On Access to Information about Government Activities" prepared in 2002 and since then circulating along the corridors of power without any tangible advance to final approval (Dmitriev, 2004).

It is noteworthy that the Russian Federation's economic and social development strategy for the period of 2000-2010 elaborated in the late 1990s by the Gref's Center of Strategic Studies included also a section on the strategy of reform of state governance. However, this section was the only (!) part of this comprehensive document, which was not made public (Zaitseva, ed. 2003:58).

The World Bank's report stresses "All earlier reform efforts (over the last 12 years) have been comprehensively blocked by the bureaucracy they were meant to reform" (World Bank, 2004:24). Perhaps most convincing evidence comes from the highest authority – the President of the Russian Federation V. Putin, who acknowledged in his Presidential address to the Federal Assembly in 2002 that the country's huge potential was blocked by the cumbersome, inert and inefficient government bureaucracy and that officialdom resisted reforms (Otechestvennyie Zapiski, 2004:111).

In many ways a similar situation, as far as "rejection" of reform is concerned, though with its own peculiarities, has existed in the Czech Republic. This can be well illustrated by both delays in implementing the civil service reform programme and by surveys of Czech civil servants and Czech citizens. Postponement of the full enforcement of the Civil service Act (adopted in 2002) till 2005, then to 2007 and then to 2009 can serve as a clear signal of non-acceptance of the reform by both politicians and top bureaucrats. One of the reform documents described it in the following way: "Current attitude of the state officials is rather illustrative of the method of "passive resistance" towards reform efforts in central state administration. This attitude can in a way be understood since neither political leadership nor top management has given any clear signal that realisation of reform measures constitutes a definitive priority of the government" (Further Advance..., 2005 point 2.3.2).

It is noteworthy that the officials in the Czech Ministry of Interior used deliberately the term "modernization", not "reform", when they developed in 1999 a concept of reform at the level of central administration. According to Pavlik, the experts of the Ministry of Interior advocated the view that the process in question could be better defined using the term 'modernisation' especially because that process is step-by-step, long-term and of evolutionary character. Unlike reform, modernisation does not affect the substance of the system as such, but it rather modifies the way that it works (Pavlik, 2000:21). This idea of modernisation was in stark contrast with recommendations made earlier by the Czech non-government National Training Foundation in a comprehensive analysis of the Czech public administration. These recommendations called for "a complete and principal qualitative transformation of the public administration" (NTF, 1998: 1-3-2). Some authors assert that in the Czech Republic one can witness "an active fight against reform proposed by EU" (Scherpereel, 2002:6).

Opposition has taken different forms. For example, reluctance of various governmental institutions and agencies to submit to the Czech Ministry of Interior proposals for devolved competencies (the former were obliged to do this within the frames of civil service reform – A.K.). In this connection Coombes remarked that transitions in Central and Eastern Europe seem "to confirm generally the extraordinary capacity of bureaucracy to withstand political turbulence" (Coombes, 2002:7).

Celebrating his first anniversary of election to the office President Putin told a group of journalists that the main lesson he had learnt was "the extreme difficulty of fighting bureaucracy" (The Observer, 8 June 2003).

8.2.2 Problems of Reform Management

As we have shown in the preceding chapters a civil service system is a complex organization made up of various elements connected by multilateral links and relations. Therefore, management of this organization and coordination between its elements is essential for effective functioning of the whole system and even more so for its reform.

Inadequate or weak management appears to be one of the reasons that delayed or misdirected advance of civil service reform in both countries. As a rule weak quality management is characterized by a badly prepared start, incorrect diagnosis, imitation instead of innovation, hidden agendas, inconsistent approach, poor coordination, inability to obtain and command resources and win internal and external support, and absence of regular monitoring and assessment.

Civil service reform in Russia appears to have lacked a proper mechanism for reform management in particular at the operational level. The Commission for Improvement of State Governance turned out to be an ad hoc structure having irregular meetings. The Civil Service Directorate in the Presidential Administration was assigned mainly coordinating and information gathering function and did not have full authority to manage the reform process. In practice the reform has been implemented by single ministries and regional authorities often duplicating each other and experiencing lack of uniform approach to reform. The weakest points of reform management were poorly defined responsibility, shortage of clearly formulated specific targets and deadlines, sporadic monitoring of implementation process and actual absence of evaluation of the reform progress and achievements. Relevant work done in this respect by international donors could not substitute for comprehensive and regular assessment that should have been done by official authority.

The overall approach to management of civil service reform in the Czech Republic evolved as more comprehensive and well thought over. The reform implementation mechanism was based on project management principles with structures and responsible persons in place and with assigned specific authority and responsibility: e.g. project teams and project managers, team leaders, reform managers in each large ministry or institution, reform steering committee and, the Office of the Government acting as the central management and supervision body.

The biggest problem for reform management in the Czech Republic appeared to be lack of sufficient support for its mission from the political authority. In practice this meant that although formally the reform was led by the Head of the Government Office, that person often could not or did not want to resolve various problems and difficulties of reform implementation, in particular issues and disputes arising between ministries and agencies.

Weak quality management can be explained also by scarcity of technical expertise in key reform areas both in Russia and the Czech Republic. There was and largely still is a severe shortage of experts with hands-on practical experience in reform design and implementation. For example, the draft Civil service Act was produced by two part-time experts who occasionally could get support from assigned functionaries representing a number of Czech ministries (Onisko, 2000:71).

The situation was further aggravated by insufficient public relations activities and other measures to promote and spread information about the reform. The weak communications resulted in a lack or poor understanding of reform objectives, principles and contents even by civil servants. Findings of a survey conducted in 2003 by the Russian Academy of Civil Service revealed in particular, that less than 40% of civil servants fully understood the objectives and essence of civil service reform (RACS, 2004: 92). Paradoxically, there were cases in the Czech Republic when some of its parliamentarians made public statements about the necessity to launch civil service reform when it was already in a full swing.

Interesting commonalities between the two countries can be found in the management of the process of reform preparation. It appears that civil service reform preparation and coordination centers in Russia and the Czech Republic were created and dissolved in a surprisingly similar manner. The Czech "Office for Legislation and Public Administration" (OLPA) established in 1992 soon after the formation of the Czech Republic was dissolved in 1996. In the Russian Federation a similar unit called Roskadry was created in 1991 (after the adoption of the first Constitution of the Russian Federation) and it was also disbanded in 1994. No sound reasoning was provided in either case explaining the cause of the closure. After liquidation the centers' competences and functions in their entirety were handed to no other organisation. Therefore, for some time civil service reform preparation in both countries lacked a central agency with the authority and political backing to exercise overall leadership.

Reform management was also affected by general weaknesses of the existing systems of public administration both in the Czech Republic and Russia, e.g. weak level of inter-ministerial coordination (Onisko 2000, Kotchegura, 2002) and shortage of qualified and well trained staff (Kotchegura 2006, SIGMA 2003).

The following example provides illustration to the overall extent of management and implementation constraints in particular in the Czech Republic. Johannsen and Nørgaard carried out a survey of former and acting ministers and central government officials in 15 post-communist countries investigating the number of policy implementation problems they experienced. In their survey they confronted 888 interviewees with 8 implementation problems and later processed the responses to find out average frequency these problems emerged in the work of officials in respective countries. The returns showed in particular that only 0.7 percent of the interviewees did not recognize any of the problems. The returns also demonstrated that the number of reported implementation problems varied from country to country. The survey results provided evidence that the state administration in the Czech Republic was in severe distress reporting an average of 4.8 problems, considerably exceeding the cross-country average. For example, the average level of problems for Armenia was 1.6 (Johannsen and Nørgaard, 2004). The survey did not cover the Russian Federation but there are firm grounds to assert that the extent of policy implementation problems in Russia were of the same order as in the Czech Republic or even greater.

Assessing the intermediary results of civil service reform the Head of the Inter-Agency Commission D.Medvedev admitted in an interview that "the reform got stuck in bureaucratic problems, in inability to organize work and activities in an appropriate manner" (Izvestia, 27 December 2005). Similarly, according to the Head of the Office of the Russian Government S. Naryshkin up to 40% of the Russian President's orders and instructions are either not fulfilled by the indicated deadline, or are not fulfilled at all (Izvestia, 25 September 2006).

8.2.3 Goal Displacement and Mislaid Priorities.

Goal displacement is a frequent phenomenon in the various types of reforms. The public administration reform in the Czech Republic provides a good example as throughout the 1990s it was regarded entirely as the process of decentralisation and reform of the territorial government. "The issue of regional reform (in the Czech Republic – AK) has become almost synonymous with administrative reform" (World Bank, 1999:5). The reform of the Czech civil service "is marginal to the reform of public administration, whereas it should be its core" (Onisko, 2000:65). "Throughout the years the administrative reform was reduced to one single aspect: decentralization, manifested by the creation of regions" (Onisko, 2000:70).

Rather common for the two countries has also been the view that civil service reform is a purely legal and organizational/technical matter aimed at enhancing efficiency of civil servants' activities. In line with such an approach: a) the reforms were reduced largely to adoption of relevant legislation; b) developing a proper mechanism of enforcement of the adopted legislation was considered of low priority; and c) the issues of civil service's mission, its place within the existing institutional setting, accountability to citizens and civil society, openness and transparency have became either irrelevant or marginal in practical terms.

The NTF report states that "the public administration reform has been understood in the Czech Republic in a very narrow way as the setting-up of regional authorities and the adoption of the Civil service Act. The aims of reform, which would secure the compatibility of our public administration with the Western level are much wider and deeper, they concern the very substance of public administration, its role, functions and means. Therefore, it will be imperative to follow the vision of the desirable future development of the Czech society, economy and state. The decisive question is: "what kind of state do we want and what role should this state play in the future development of our society and country?" (NTF, 1998 point 7-1-7).

In Russia the civil service reform concept and programme left the issues of consistent and targeted fight against corruption virtually unaddressed. It is true that the new Law 'On State Civil Service of the Russian Federation" introduced a concept of conflict of interest. Moreover, even commissions on the regulation of conflict of interest were set up in the majority of ministries. However, in the course of at least two years since the adoption of the law in 2004 not a single case has been reviewed by these commissions. All this was happening against the background of numerous reports on rocketing corruption in the state bureaucracy.

Another example of wrongly selected priorities is manifested in the clear underestimation of the significance of democratization of the state administrative apparatus, above all in terms of making it more transparent and accountable externally. In fact, if we examine recently adopted Czech and Russian legislation on civil service we may conclude that one can find there very little concerning accountability of the civil service to citizens and society. Rather, there are non-specific references to the need for ensuring transparency of the service's activities.

8.2.4 Inadequate Preparation of Reform

Were the reform proposals firmly based on prior studies of the problems facing state administrations in Russia and the Czech Republic? If consider the period until the late 1990s the answer is rather negative.

The report prepared by the Czech National Training Foundation stated "No real analysis and evaluation of the development of the state administration from the first republic up to the present has been made yet" (NTF, 1998:12). Another Czech source explained "One might think that elaboration of such an important law defining status of civil servants would have been preceded by sound quantitative and qualitative analysis of the state administrative personnel and comparative study of different analogous systems abroad, as well as by the development of various alternative models. In reality, this has not been done" (Onisko, 2000: 65).

In essence preparation of civil service reform in the Czech Republic was carried out in a confidential manner. It was marginalized within the framework of public administration reform whereas it should have been its "heart". The reform lacked strategy and the authors of the law progressed in "complete darkness" (Onisko, 2000:65).

A similar situation in this respect existed in the Russian Federation at least throughout the 1990s. The preparation of the Law 'On the Basic Principles of Civil

Service of the Russian Federation' (adopted in 1995) was not preceded by a comprehensive study of the Russian state administration. Although the law was in many respects a positive step forward, it did not tackle in a consistent and robust manner some important issues of the development of the Russian state administration. Elaboration of civil service reform concept and programme was also done "behind closed doors".

Good quality assessment and analysis regarding status and problems of the Czech and Russian civil services finally appeared in the late 1990s. Conceptual documents prepared in Russia by President Yeltsin's advisors in 1998 and in the Czech Republic by the National Training Foundation in the same year serve as good examples. However, these reports were not used at that time and were largely disregarded in the course of later preparation of reform programmes and legislation.

The same can be often said about recommendations given by donor agencies. For example, in 1995 SIGMA provided a detailed commentary to the draft Czech Civil service Act. But its conclusions were largely neglected (World Bank, 1999:7).

Typical of reform implementation in both countries was underestimation of the need for public support of reform which is one of the key factors of its success.

In Russia the text of the concept of civil service reform was not made available in full to the public until almost a year after its approval by the President. Similarly in the Czech Republic neither citizens and NGOs, nor rank and file civil servants were invited to any significant degree to take part in discussing draft concept and other important documents of the forthcoming civil service reform. One of the documents prepared by a Phare project team put it this way: "Public servants view the reform as a purely internal matter, whereas the public (society) either does not understand or does not know of the benefits which the reform may bring" (Phare, 2000).

Generally preparation and execution of reform were not always well thought through. It is widely acknowledged that profound restructuring of the Russian Government in 2004 was carried out in line with international practice. But as a World Bank report notes it was "a very radical and risky reform, which will be extremely difficult to operate, particularly since it was launched with no advance consultation or communication. Even inside the system, most participants do not fully understand how it is all supposed to work in the future. Shifting boxes in structure charts – no matter how radically – does not by itself change the behaviour of individual civil servants and institutions. There have to be new incentive and accountability systems, and procedures for translating government restructuring into real administrative reform in order to achieve real improvement in performance of the civil service." (World Bank, 2004:17).

8.2.5 Corruption

Corruption (no matter which form it takes) undermines the goals of civil service reforms and the mechanism of their implementation. The reform which does not address the issues of eradicating conditions for corruption, embezzlement and fraud runs high risk of failure. Raising the effectiveness and efficiency of activities of corrupt civil servants very often contributes to widening opportunities for their personal enrichment.

Corruption is acknowledged to be a very serious problem in Russia. It is estimated that annual corruption in the state administration amounts up to USD 33 billion (INDEM, 2002). So far, adopted anti-corruption measures have been ineffective. The Anti-corruption council set up by President Putin in 2003 has not yet shown any activity. Existing anti-corruption legislation is often ignored in practice. For instance, despite the mandatory requirement of holding open tenders in public procurement, only 51% of the state budget funds allocated for that purpose in 2001 were distributed as a result of open tenders. (Nikolaev, 2006 :136).

As compared to Russia the scale of corruption in the Czech Republic does not seem to be so extensive but according to a SIGMA report "corruption still remains a significant problem, having penetrated the majority of public administration structures at both central and regional levels" (SIGMA, 2003:13). The European Commission stated in one of its reports that "surveys of public opinion (in the Czech Republic – A.K.) show a consistent increase in the perception of corruption and economic crime. Concern is greatest as regards the state administration, the police and intelligence services, healthcare, banking and the political sphere" (Regular report..., 2001: 20).

According to the results of a public opinion poll conducted in May 2001 by Gfk Institute (Prague), half of the population of the Czech Republic agree with the opinion "we live in a corrupt state" (GfK, 2001).

A fairly remarkable illustration and an interesting similarity between Russia and the Czech Republic is provided by the World Bank study of the influence of corruption in state administration on businesses in CEE and CIS countries. It shows that the share of firms' annual revenues spent on bribing officials in Russia and the Czech Republic is almost equal at about on average 4.5% of the companies' annual income.

The Czech Republic and Russia saw repeated scandals over party and election financing. In 1996 prior to the Presidential elections, the security of the White House (the building housing the Russian Government) intercepted five hundred thousand US dollars in cash brought by a representative of one of the Russian banks to the managers of President Yeltsin's election campaign. As it turned out later the representative received a transaction of this amount in the Russian Ministry of Finance.

In the Czech Republic the ruling party, the ODS, could not explain: why a dead Hungarian and a resident of Mauritius were its main donors during 1995-1996; why it concealed over 170 million Czech crowns in Switzerland; or why it received 7.5 million crowns from Moravia Steel Plant after the company was allowed to lower its bid for Trinecke Zelezarne Metal Works by 300 million Czech crowns (Grzymala-Busse, 2001: 46).

In the opinion of a prominent academic expert, today in Russia one can see a poorly controlled conglomerate of government agencies which has merged with big business clans in order to develop their joint financial/bureaucratic interests. As President Putin said once sarcastically, "the depths of the government are as deep as the oil and gas reserves of the Russian Federation and it is certainly true that things sometimes vanish there." (Arbatov, 2007).

Overall it can be stated that corruption considerably undermined public confidence in civil service reform processes in both Russia and the Czech Republic.

8.2.6 Clientelism and Patronage

Clientelism denotes such social relations where personal loyalty to the patron prevails over democratic decision-making, professional duties and ethical behaviour. Our analysis has shown that clientelistic networks (widespread in the former socialist countries) survived political upheavals and remain the base of relations across state administrative institutions in many post-communist societies. "In Eastern Europe, clientelism - in interaction with various forms and levels of corruption – is becoming a stable form of social organization" (Sajo, 1998).

Within such a framework qualifications, skills and competence have less significance than personal loyalty to the boss when it comes to considering promotion or increased remuneration. Furthermore, particularly in Russia but also in the Czech Republic, most top level civil servants are members of informal "teams" based on political, personal, place of origin, joint studies, family and other affiliations (Afanasyev, 2004). The teams safeguard the individual interests of the team "leaders" and those loyal to them within administration. If an individual wishes to gain promotion and be successful in the system he should become attached to someone in a higher position: join the team and serve the group's interests.

Such a system is maintained also by existing practices and regulations. For example, various bonuses constitute a considerable part of pay arrangements in both Russian and Czech civil services and are almost entirely at the discretion of managers.

Widespread clientelism creates a situation when officials have little or no incentives to safeguard interests of the state and citizens or promote competition in various spheres including career appointments. Instead, they establish personal relationships that are beneficial for individuals and separate companies but detrimental to society at large (Linkola and Luhtala, 2001:5). Such informal networks and practices are very often linked to vested interests, especially of those who profited from illegal privatization. In turn, these networks create fertile ground for politicization of civil services.

According to Afanasyev, clientelistic networks have penetrated all layers of the Russian bureaucratic officialdom and determine to a large extent its principal characteristics (Afanasyev, 2004:16). Similarly in the Czech Republic "despite the liberal rhetoric of the past decade current Czech society is to a very large extent a clientelistic and corporate one. Groups of friends, acquaintances and relatives dominate most of the public sector, including the administration, media as well as public schools and academia" (Tychtl, 1999).

This assertion is confirmed by Nunberg who wrote "In the main, administrative practice from the communist era has survived remarkably undisturbed, though, by and large, in a de-ideologized form" (Nunberg, 2000:254).

A number of prominent experts and scholars hold the view that in the Russian Federation and the Czech Republic the political elite possesses some principal features of the "nomenclature" (a communist type bureaucracy) – above all, a high degree of disregard for the society. Many of them speak about "clientelism and pressure groups usurping the right to adjust the interpretation and implementation of laws, especially subordinate norms, to serve their own benefit". (Potucek, 2003:48). Some experts conclude that political and administrative elites in Russia and the Czech Republic have merged into *"le nouveau nomenclature"*. A prominent Czech philosopher Vaclav Belohradsky points to "an arrogant new political and economic nomenklatura" that was formed in the Czech Republic in the nineties (Central Europe Review, No.20, 2000).

8.2.7 Mentality and Administrative Culture

The importance of studying national mentality and administrative traditions can hardly be overestimated. "Behavioral traits, more than structural elements, are what distinguish one bureaucracy from another." (Heady, 2001).

Dominant mentality and administrative culture (attitude towards authority and citizens, prevailing qualities, values) in Russia and Czechia seem to have much in common since their roots stem from the heritage of a pre-communist monarchist past (Russian and Austrian-Hungarian Empires) and subsequent communist regimes. It is true that Czechoslovakia went through a period of democracy in pre-second world war years but the period seems to have been rather too short to have a long-term impact.

Among frequently mentioned typical national features are a high decree of conformism of the Czechs (reflected in such well-known character as Hasek's Good Soldier Sveik) paralleled by generally low public demands and high degree of patience of the Russians.

There still remains a basic principle of social culture and mentality widespread in the Russian Federation and the Czech Republic which is based on the view that survival is possible only by offering mutual favours. Such attitude feeds patronage and clientelism, the roots of which can be traced back to the times of the 19th century monarchies ruling over the territory of both countries.

Mass consciousness in the Czech Republic and Russia suffered a sort of psychological crisis in the 1990s. This was connected with feelings of lost identity (dissolution of the Soviet Union and Czechoslovakia) and realization that previously the most developed country in Eastern Europe (Czechoslovakia) and a world superpower (Soviet Union, later Russia) could no longer claim their former statuses.

Indeed, Czechoslovakia was a leading industrial state both during pre-war period and after the war with a relatively high standard of living as compared to other communist states. At present the Czech Republic is only one of several Central European nations often lagging behind others in some key areas of social, economic and political development, affected by serious economic decline in the second half of the 1990s and until recently (before accession) criticized frequently by EU officials.

Such a situation is even more typical of Russia since the comparison of economic, scientific, human resources, military and other potential of the Soviet Union and that of present day Russian Federation reveals distortions comparable only to the harmful consequences of devastating wars.

There are of course peculiarities inherent to national mentality in each of the countries in question. For instance, the Russian Orthodox church occupies an influential position in society with millions of followers. At variance to many other post-communist countries, Czech mentality is distinguished by weak religious feelings and relative immunity to ultra-nationalism.

At present there does not seem to be a nationwide or other idea (clerical, nationalistic, charismatic, cultural etc.) that unites either Russian or Czech societies.

The typical features of the prevailing bureaucratic culture in Russia are demonstrated in the results of the sociological interview of Russian civil servants conducted in 2003. When asked what type of negative qualities are most typical of Russian civil servants today, the following answers were received (ranked according to their extent): a) red tape; b) use of official position for personal gains; and c) imitation of activities (RACS, 2004 : 48).

Commenting on the Czech state administration, Sean Hanley describes it (perhaps too emotionally) as the combination of Austro-Hungarian and Soviet type of bureaucracy "melted in one pot", which produced a strange cocktail. "In fact, a more bureaucratic, bureaucratically minded nation can scarcely be imagined....the plain truth is that bureaucracy is as Czech as beer, Milan Kundera and Good Soldier Sveik" (Hanley, 1999).

It is interesting that a survey by the Czech Academy of Sciences conducted in 2004 and based on interviews with civil servants from the Czech Republic and other EU member countries' revealed that whereas the West European civil servants try to work efficiently and confront problems as they arise, "many Czech officials try to avoid a problem-solution approach. What they prefer to do is impute guilt on a third party" (Radio Prague, 31 March 2004).

Similarly in Russia "many civil servants' actions are directed not to achieve a result, but to coordinate and agree the form and contents of the document or decision with other bureaucrats" (Otechestvennyie Zapiski, 2004:86). Furthermore, recent years have provided enough evidence for the claim that there exist large groups of civil servants in Russia who regard their service and duties entirely from the point of view of rent seeking opportunities.

Overall it may be stated that the existing administrative culture in Russia and the Czech Republic regards secrecy and personal loyalty as a virtue. Responsiveness, transparency, responsibility and dialogue with citizens are viewed as a rule as matters of secondary importance.

8.3 Extra Bureaucracy Factors

8.3.1 Politicization of Civil Service

Proper development of the institution of civil service in Central and Eastern Europe faces a serious challenge of overcoming the custom of treating state administration as the instrument of personal or narrow group power. The habit of placing „their own people" in the strategic administrative positions and to exercise meticulous control over high ranking civil servants has been a long tradition in the administrative cultures of Central and Eastern European ruling elites and has largely survived until today.

Indeed, recent experience shows that in many post-communist countries the political parties that won elections offer public service positions to their supporters and provide benefits to their collaborators in the form of access to various government posts and services. Far from exceptional are situations when the new governments act as clients of the patrons who helped the governing parties win elections.

The relationship and interplay between new political elites and the predominantly unchanged bureaucratic establishment exert serious influence on the process and results of civil service reform. Observation that "bureaucracy plays a distinctly political role during regime transitions" (Church, 1981) made with regard to developing countries, in our view, is highly relevant to the situation in Central and Eastern Europe and CIS countries.

As a rule, change in the relationship between top civil servants on the one hand and politicians on the other, is most difficult because "numerous interconnected features of most bureaucracies require not piecemeal change but a revolutionary restructuring of the whole system to make major differences, and, because of vested interest, major restructuring will be resisted and piecemeal change will often be neutralized to continue to protect the interest of the bureaucrats" (Chandler, 2000: 258).

The high level of politicization of state administration, absence in practice of standard criteria for promotion and benefits and dependency on individual personal ties contribute to maintaining instability within the service and shape the conduct, attitudes, values and objectives of politicians and civil servants in both countries.

In the Czech Republic legally defined distinction between political appointments and career civil servants has been absent for a long time. "There is no clear distinction between political appointments and career positions. In case of political changes all the top administrative cadre is under threat of losing their posts." (Onisko, 2000:21). Although in Russia such a distinction was formally introduced in 1995 by the Law 'On the Basic Principles of Civil Service of the Russian Federation', in practice this distinction was not always clear. Civil servants' careers have been particularly vulnerable at the regional level where changes of the political leadership were more frequent and the scale of replacement of the civil service staff considerably greater. Instances of full scale reshuffle of the top and mid level civil servants upon arrival of a new Governor or Minister are so frequent that its absence may be considered exceptional.

"The system (in the Czech Republic - AK) allows politicians to create political clienteles within the administration in an unrestricted way, as there are no legally established limits for political involvement of civil servants" (SIGMA, 2003:10). "All employees (in the Czech state administration –AK) appear to work under a cloud of uncertainty regarding the security of their posts, which seems to increase after major political changes and to be greater the higher one's position" (World Bank, 1999:81). This is particularly true with regard to top level officials, who are often replaced after each elections (Onisko, 2000:21).

The situation in Russia in this respect is no better as compared to the Czech Republic. For example, there have been numerous cases of using the so called "administrative resource" by the leading political parties during the elections campaigns both at the federal and regional levels. Illustrative in this sense is the new law 'On the State Civil Service of the Russian Federation' adopted in 2004. Contrary to the provisions of the previous law (adopted in 1995) the text of the new law does not contain the requirement of de-partisanship of the civil service.

In connection with this quite interesting are the results of a survey of the Czech and Russian civil servants conducted in the Czech Republic and Russia. The survey was carried out in the Czech Republic by R. King (King, 2001) and in the Russian Federation by the author of this study using the same questionnaire (in reduced form in Russia – A.K.). Nineteen Russian and nineteen Czech civil servants were interviewed representing various government ministries and agencies at central and regional level. Around 85 percent of interviewees had more than five years of service record. The findings of the survey reveal high level of politicization of both Russian and Czech civil services – over 90% of the responses point clearly to high politicization of both civil services.

The following tables provide data on the responses to a number of key questions of the survey:

In your country civil service and politics are kept separate?	Russian Federation	Czech Republic
Yes	2	-
No	17	19

In your country change in Government entails replacement of some career civil servants?	Russian Federation	Czech Republic
Yes	17	19
No	2	-

Protection of civil service from direct political interference is	Russian Federation	Czech Republic
Not important	-	-
Important	11	-
Very important	8	19

It should be noted that whereas in the Czech Republic interviews were held exclusively with civil servants with over 5 years record of service, in Russia around 10 % of interviewees possessed less than 3 years of work experience. This can be one of the reasons for slight deviation in responses of the Russian civil servants as compared to their Czech colleagues.

8.3.2 Lack of Political Will

It is widely acknowledged that the Russian and Czech political leadership demonstrated lack of political will to reform their civil services throughout most of the 1990s. Various reasons can be offered to explain this fact: for example, adherence of Prime Minister V.Klaus and some leading Russian liberal economists to the theory of reducing the state to minimum "since market will put everything in order"; or prevailing mentality and soviet-style conception that the state is for their ownership. These reasons can be generally accepted but only as partial explanation.

As we have shown in previous chapters one of the principal causes of this unwillingness derives from the top politicians' fear of losing control over the bureaucratic establishment and undermining their client type relationship with the state administration's elite. In many ways a remarkable statement was made on this issue by a Czech senator Edvard Outrata. Speaking in the Czech Senate on the question of enforcement of the new Civil service Act he said "The civil service is the part of our transformation process that has not yet been completed, and I see a kind of conspiracy running from one end of the political spectrum to the other not to finish it. Do you politicians really believe that you're better off if you can manipulate the personnel in your offices? I tell you that you're far worse off, because you won't get expert help where you need it; you'll get help from your buddies who've already long ago told you everything they have to say...We are returning to the level we thought we'd escaped in 1989. Do we really intend to remain stuck there?" (Wilson, 2005).

In the case of the Czech Republic it appears that the political will has been absent throughout all the country's post-communist period. The decision to launch civil service reform and adopt a Civil service Act was taken under consistent and heavy pressure from the European Commission. One of the public administration reform experts used the expression "battering ram" to describe the pressure from EU on the Czech Republic on the issue of civil service reform (Behlavy. 2002:64). It is interesting that after the Czech Republic joined the European Union the Czech authorities seem to have lost all interest in successful implementation of the civil service reform programme.

Similarly in Russia, "patron-client" relations penetrating bureaucratic officialdom constituted a de-motivating factor for any action aimed at modernisation of the state administration (Afanasyev, 1998).

The political will to launch civil service reform has emerged only following the election of the new President, who committed himself to transform an inefficient and unaccountable state administration into a modern civil service. President Putin did not seem to be "fascinated" by the existing patronage networks. Many of them were cementing the "family" of the "old guard" as was made clear from the opposition to his new course. Speaking about reasons that made civil service reform possible in Russia Gray noted that "the election of Putin...provided a window of opportunity for change which appeared to have been taken in the early part of 2002" (Gray, 2004:14).

8.3.3 Budgetary Constrains and New Opportunities

As we have already pointed out in Chapter II, contrary to the experience of many industrialized countries where civil service reforms were driven among other things by economic considerations, namely, intentions to reduce the cost of maintaining huge state bureaucracy, budgetary constraints in the post-communist countries appear to have impeded rather than encouraged reform initiation and implementation. Civil service reform overall is a quite expensive undertaking, and is particularly so for countries involved in large scale transformation and experiencing severe budget deficit. It is necessary also to bear in mind that administrative reforms in CEE countries are aimed at reconstructing the existing administrative system, not simply modernizing it. "So, the question was not only, or even primarily, one of downsizing, but rather one of building, instead of reforming, a functioning PA system, which is invariably costly" (Verheijen, 2000:41).

Therefore, fiscal constrains could be an important factor that reduces capacity for civil service reform in the post-communist countries. This factor played a definite role in the case of Russian and Czech reforms.

It is noteworthy that the first draft Civil Service Law submitted to the Czech Government in 1993 and revised many times in 1994 granted civil servants considerable privileges and was strongly criticized by the Czech Ministry of Finance. This draft was finally abandoned in 1995 with explanations of budgetary restrictions (Onisko 2000:12). In explaining the decision to postpone full enforcement of the Civil service Act (adopted in 2002) in the Czech Republic, reference was also made to the shortage of funds.

Insufficient funding of state administration resulted in situations when, for example, vacancies for some key positions in the Czech Institute of Public Administration (created in 2000) have been left open for several years. Lack of interest from potential candidates was caused mainly by the low compensation package offered.

Throughout the 1990s the situation in Russia in this respect was even more deplorable. The state budget deficit and low priority attached to the conditions of employment of civil servants resulted in a situation when wages of Russian civil servants were unacceptably low especially as compared to the wages in the commercial sector. Low level of funding hindered hire of qualified and experienced local and foreign experts to assist reform preparation and implementation.

Although fiscal constraints can not be regarded as the main factor explaining delays and difficulties in launching and implementing civil service reform (there are examples of other reforms implemented at that time in spite of severe budget deficit) they constituted one of the obstacles on the way of reform preparation and advance.

It should be noted that the far from exemplary overall economic situation in both countries in particular in the second half of the 1990s has cardinally changed in the early 2000s in Russia. The main reason – steep increase in the price of oil on the world markets. Russia has benefited from this trend and improved considerably its financial position. For Russia this period was marked by high rate of economic growth and larger revenues of the state budget.

With its budget benefit in the years 2000–2005 running at the average rate of about 4-5% Russia differed considerably from the Czech Republic where the deficit of the state budget has become permanent. However, low effectiveness of the budget allocations and high level of corruption have prevented Russia from using most efficiently this "advantage" (compared to the Czech Republic) at least in terms of advance of civil service reform. Furthermore, the growing flow of petro-dollars seems to have caused a number of unanticipated side effects. The most striking are the steep increase in the expenditures on the Russian state administration and growth of the number of civil and municipal servants.

Experts of Troyka Dialogue consulting company have calculated that during the period 2002–2006 labour productivity in almost all sectors of the Russian economy was rising (e.g. growth in agriculture and construction up to 10-15 %); only in the energy and financial sectors it remained static. But labour productivity in the Russian state administration has fallen 7 % during the same period (Business Week Russia, 28 August 2006).

From the year 2000 till 2006 overall expenditure on the Russian state administration has increased 11 times in nominal prices. For the same period growth of allocations to the social sphere in nominal prices was only three fold. Draft state budget for the year 2007 envisaged further 50% (!) increase of budget spending on the Russian state bureaucracy (Business Week Russia, 28 August 2006).

The Russian Minister of Finance A.Kudrin provided the following data – in 2000 an increase in one dollar of the oil price on the world market led to 0.2% growth of Russia's GDP, but in 2005 the same increase resulted in only 0.006% GDP growth. During the same period the number of civil and municipal servants in Russia had gone up 1.3 times and budget expenditure to support their existence had grown from 0.3% GDP to 1.1% GDP (ibid).

These data cannot fail to illustrate an interesting tendency typical at least of the early 2000s - the higher world oil prices and hence revenues of the state budget were, the more Russian state bureaucracy would expand and consume resources. We may also conclude here that the civil service reform in Russia has failed so far to reverse this trend.

8.3.4 Constrained Political Competition

It is commonly recognized that one of the effective instruments for reducing politicization and enhancing accountability of state administration is *critical opposition*: strong opposition parties that continually threaten to democratically replace the government in power. Furthermore, critical opposition tries to force the ruling party/coalition to be consistent in implementing its commitments and pledges, and to effectively and timely respond to new challenges. Experience shows that those political leaders who lack political will to implement required policies often change their mind under pressure from critical opposition which is using various instruments (parliament, mass media, non-governmental organizations) to encourage appropriate decision-making. More importantly the existence of strong opposition reminds politicians of the inevitable accountability for the policies they pursue.

"When no critical opposition monitors the government, and forces the creation of formal oversight and power-sharing institutions, governing parties can freely dispense state funds, jobs, contracts, and other resources for their own private benefit" (Grzymala-Busse, 2003).

Such a situation (no critical opposition) developed gradually in the Russian Federation and the Czech Republic in the early 1990s. In Russia the turning point to elimination of effective opposition took place in 1993 when President Yeltsin ordered armed assault of the building of the Russian Parliament. Following this violent suppression of the opposition, its role since then has been limited and depends largely on the balance of political forces in the country and political preferences of the President in office.

The situation in Russia is notably different also because the results of parliamentary elections have no direct impact on the formation of the government. The Prime-Minister is appointed by the President (later to be approved by the Duma) who is not strictly obliged to take into account the outcome of the elections in selecting candidates for this post. The same rule applies to future ministers who may not be chosen only from the list of representatives of the winning party/coalition. Besides, the system of political parties in Russia continues to remain largely in embryonic state.

Opportunities for fare political competition were reduced further when in the early 2000s the Duma adopted a number of laws which established much stricter criteria for the parties wishing to be registered for elections. This move made it practically impossible for small and mid-size political parties to take part in parliamentary elections.

In the Czech Republic the almost uncontested hegemony of the Civic Democratic Party – ODS (1993-1996) made it difficult to foster a genuine pluralist competition with clear governmental alternatives. The Social Democratic party, as ODS successors were hardly different in this respect The Czech political party system in consequence resembled an incomplete and somewhat "defective" pluralism (Akmiska, 2000).

The period starting from 1997 was marked by a formal deal between two major parties. The so called "opposition" agreement reached between the Social Democrats and ODS (headed by Klaus) in 1997 stipulated that whichever of the two parties had won an election, the other would not support a vote of non-confidence in parliament. This agreement between the biggest right and left wing parties was also aimed at squeezing out the smaller parties from competition.

As we can see the intentions to ensure a sort of "monopoly" position during the elections was typical of both Russian and Czech largest political parties.

In essence the Czech Prime Minister V. Klaus and the Russian President B. Yeltsin (who both, as we indicated above, were reluctant to start genuine reform of state administration) were under no serious pressure (that should usually come from the political opposition and parliament) to modernize and reform the existing administrative system and bureaucratic apparatus. "Political culture of the country has suffered in the last four years because the ruling coalition governed virtually unopposed. The ODS leaders and Klaus in particular, have tended to disregard or show disrespect for not only the opposition and the media but frequently to their coalition allies as well" (Pehe,1996 : 28).

We may conclude that seriously weakened competition between major political parties and forces representing various political spectra in the Czech Republic and Russia was a significant factor that allowed delays in launching civil service reform in these countries happen.

8.3.5 Alienation of Citizens and Weak Civil Society

High level of alienation of citizens from the state administration and political leaders is one of the factors that feeds growing indifference and lack of motivation of citizens to engage in constructive criticism and participatory action.

The Czech President V. Gavel speaking at the joint session of the Chambers of the Czech Parliament in December 1997 stressed that "growing number of citizens feel disgust towards politics and politicians, whom they rightly blame for recent failures" (Shepherd, 2000:47). In Russia the state administration, freed from former control by the communist party, in fact distanced itself further from society (Obolonsky, 1999:308).

A paradoxical situation has arisen where new political elites have not developed respect for their own society. This lack of respect was occasionally manifested in concrete actions and contributed to mutual mistrust and the difficulty of the entire reform project. As an example one can refer to the separation of the

Soviet Union and division of Czechoslovakia which appear to have been accomplished against the will of the majority of citizens. According to sociological polls in March 1992 only 17% of the Czechs and 11% of the Slovaks supported the idea of division of the country into two separate states. The new Czech and Slovak elites "decided not to submit the deal (separation) to a popular referendum fearing possible rejection" (Shepherd, 2000:138).

Similarly the leaders of three largest Soviet republics (Yeltsin, Kravchuk, Shushkevitch) adopted hastily a decision on dissolution of the Soviet Union in December 1991. They had rejected the idea of holding a referendum and ignored the results of referendum of March 1991 which clearly demonstrated that absolute majority of the population was against break up of the Soviet Union.

The study providing results of the polls conducted by the Centre for Public Opinion Research at the Institute of Sociology of the Czech Academy of Sciences shows that the national elites in the Czech Republic are evaluated very critically by the public and this critical appraisal has only intensified in recent years.

"Over the course of the 1990s there occurred a decline in the amount of confidence people have in both political parties and top politicians, and in how politics functions on the whole in addressing society-wide problems and asserting the valid interests and needs of ordinary people. The majority of the public evaluates the political culture and morality of the highest political representation, including its motivations, in very critical terms. This negative evaluation of politics in general and politicians in particular manifests itself in the decline in interest in public affairs and undoubtedly also in the falling participation in elections" (Mišovič and Tuček, 2003:11).

In general one of the peculiarities of the Russian and Czech transition was the absence of a long term and permanent interaction between the "new" political elite and the society at large.

The factor of generally weak civil society in both Russia and the Czech Republic was clearly hindering the progress of civil service reforms. "The best guarantee for an improved public service is not internal administrative reforms but a stronger and more vibrant civil society that can hold public officials accountable" (Pierre 2001:181). The role of civil society was underestimated by Prime Minister Klaus and almost completely disregarded by President Yeltsin and the Russian political elite as a whole.

Among key factors that contributed to distancing civil society from the state were corruptive behavior and the arbitrariness of government officials. In Russia the distance between civil society and the state administration is reported to have been steadily growing throughout the 1990s and later (Nozdrachev 1999, Obolonsky 1999, Larjavaara, 2001, Kotchegura 2004).

Some experts reckon, in our view justifiably, that civil societies in Russia and the Czech Republic are still in a way "infantile", meaning they are not ready to face responsibility. Instead they tend to look up to "charismatic" leaders from whom they derive self-confidence and certainty, the way children derive their self-confidence and certainty from their parents.

8.3.6 Historical Legacy

Commonalities in history between Russia and Czechia go back to the very early days of statehood in both countries, which geographically were parts of a transitional zone between the East and the West. Both nations were under foreign domination for several centuries. However, Russia achieved independence in the 16th century whereas there had not been an independent Czechoslovakian state until the end of World War I. In Russia the state considerably pre-dates democracy. In Czechoslovakia democracy and national state emerged almost simultaneously.

There are well known natural, geographical, demographical and other differences between the Russian Federation and the Czech Republic. One can simply compare, for example, the huge largely unpopulated territory inhabited by multinational multi-religion ethnos (Russia) and a relatively small single nation country (Czech Republic). On the other hand, the Russians (who make the majority of the population in the Russian Federation) and the Czechs belong to the same ethnical group – the Slavs.

Both countries have a set of similarities explained by their recent communist past. A Soviet model of governance with slight variations was dominant in all former socialist countries. Its characteristics were the leading role adopted by the communist party, a socialist form of property ownership, merger of the top party and state officials into one ruling elite (nomenklatura), and democratic centralism as the key principle of public management.

The spirit of Prague Spring with its aspirations for democratic and human values is still fresh in the memory of many Czechs and that makes a distinctive difference from the situation in Russia. Before the start of Gorbachev's perestroika, Russia knew only a very limited period of democratic development which lasted six months after the bourgeois revolution in February 1918 to the Bolsheviks' takeover in October the same year.

But one should also remember that after suppression of the Prague Spring in 1968 the regime in Czechoslovakia became one of the most rigid and orthodox in the communist block and there were no widespread opposition movements in the country similar to Solidarity in Poland. The same can be said about Russia where the dissident movement was represented by a small group of citizens.

By late 1991, both in Russia and in the Czech Republic, the supremacy of the communist party was abolished, competitive elections were fairly organized and held, and a form of political pluralism started to take root.

8.3.7 Constitutional Setting

The extent and balance of powers belonging to the President, Parliament and Government in the two countries proceed from the peculiarities of the form of state organization, political traditions and historical development. The Russian Federation is a parliamentary federal republic with a strong presidency and the Czech Republic is a unitary parliamentary republic but with rather limited powers of the president.

It is noteworthy that there is certain dualism of the executive power in the Russian Federation since in practice it is exercised by both the President and the Government. The government in the Czech Republic is formed by the party or coalition that has won the elections whereas in the Russian Federation the results of the elections to the Duma and formation of the government do not necessarily co-relate. "The major drawback of the Russian government is that it is purely bureaucratic i.e. it is not formed by the winning party or coalition" (Delyagin, 2005). Table 8.1 demonstrates to whom belongs the authority for taking important political decisions in the Russian Federation and the Czech Republic.

Table 8.1 Authority for Taking Important Political Decisions in Russia and the Czech Republic

Political decisions and powers	**Russian Federation** (reference to the text of the Constitution)	**Czech Republic** (reference to the text of the Constitution)
Dismissal of the Government	By the Presidential decree or by the Duma through vote of non-confidence (Art. 117)	Only by resolution of the Parliament (Art. 72, 73)
Dissolution of the Parliament	By decision of the President as the Head of State (Art. 109)	By decision of the President as the Head of State (Art. 35)
Possibility of combining posts of a minister and a deputy in a parliament	Not possible (Art. 97)	Possible (Art. 32) with some restrictions re. membership in the Parliamentary committees
Rule of countersignature	No such rule.	Some of the Presidential decrees must be signed by the Chairman of the Government or designated Minister
Relations of President with Government	President is empowered to chair meetings of the Government, to request information.(Art. 83)	President is empowered to take part in meetings of the Government, to request information.(Art. 64)
Presidential elections	Nation-wide elections (Art. 81)	Elected by the Parliament at the joint session of both Chambers of the Parliament (Art. 54)
Presidential right of veto	Yes, but can be overturned by 2/3 of votes in the Duma (Art 107)	Yes, but can be overturned by simple majority of votes (Art. 50)
Presidential right to initiate legislation	Yes (Art. 84)	No specific reference in the Constitution
Decision to declare war or state of emergency	By the President (Art. 87-88)	By the Parliament (Ch1, Art. 43)
Approval of the new government and its programme	Needs both approval of the Parliament and of the President	Approval by the Parliament (Art.68)
Appointment of the Prime-Minister	By the President (Art. 83) to be later approved by the Duma	By the President (Art.68) to be approved by the vote of confidence in the Parliament

The above table shows that the Czech parliament generally has more powers than the Russian Duma. The Russian President has more authority than his Czech counterpart. In the Czech Republic almost all organizational and many other issues regarding state administration must be decided by a formal parliamentary act whereas in the Russian Federation this can be often resolved by the Presidential decree. From this point of view policy decision making in the Czech Republic as a rule is more protracted.

The executive branch in both Russia and the Czech Republic appear to be quite influential. "We found an overall dominance of the executive as decision centers and that the parliaments played only a limited role – in particular in presidential systems" (Johansen, 2004 : 9).

The Czech Republic differs from Russia also as far as the status and rights of ministers are concerned. In Prague they have had the right to be simultaneously deputies of the Parliament. "It is quite usual that ministers are simultaneously deputies in the parliament – this gives them immunity, which protects them forever" (Konopasek, 2002:22).

A noteworthy distinction is that the government in the Czech Republic is responsible mainly to the Parliament whereas in the Russian Federation the government in essence is more accountable to the President. In view of the general weakness of political parties and domination of pro-presidential lobby in the Russian parliament, the nature of government in Russia tends to be more *majoritarian*. In the Czech Republic one may speak about the executive being more *consensual*.

The Constitutions in the Russian Federation (1991) and the Czech Republic (1992) resulted from negotiations between various political forces and groupings. However, the current Constitution (1993) in Russia was adopted as a result of violent ejection by the then President of one of the groups from the political arena.

It is interesting that two major political parties in the Czech Republic, the ODS and CSSD, attempted in 2001 to change the Czech Constitution and the electoral law to secure the two parties' advantage in the next elections by "squeezing out" from the election process smaller parties, but so far have failed to do so (the Senate rejected the bill). A similar attempt was more successful in Russia in 2000 when the Duma introduced much stricter criteria for parties wishing to take part in the elections.

8.3.8 Political Framework

The political and administrative systems in the former communist countries provided an example of the development of uncontrolled and unaccountable party and state bureaucracy. The political regime in Czechoslovakia before the "velvet revolution" was one of the most conservative and ideologically rigid in Eastern Europe. In this respect it was very similar to the political regime in the Soviet Union. In our view this similarity is one of the key factors that explain many common traits of civil service development and reform in both countries.

Following break down of the communist regimes and subsequent dissolution of the Soviet Union and Czechoslovakia the new political elite in these countries faced the task of creating new political systems with new institutions and functions. A list of immediate priorities included among other things: appropriate distribution of state power; constructing relationships between legislative, executive and judicial branches; establishment of proper representation of social and regional interests; ensuring political stability and interface between the political elite and the society at large; and last but not least – creation of a new system of governance and formation of a new state administration.

The reality showed that, as Coombes observed, "In most instances politicians and administrators in the countries concerned (Central and Eastern Europe – A.K.) have stubbornly refused to understand public administration in the more realistic but nevertheless rarely articulated western sense, as involving the making of decisions both politically determined and politically consequential (Coombes, 2004).

Boone and Horder described the prevailing in CEE countries tendencies in the following way:"In the vacuum that followed political breakdown, the old elites and rent seekers captured the political initiative in these countries. To sustain their powers and sequester incomes, they issued credits and maintained distorting policies and, as a result, acquired enormous assets" (Boone and Horder, 1998).

The persistence of administrative and bureaucratic practices typical of the previous regimes has been complemented by preservation of generally high level of politicization of the state administration in these countries. The situation was further complicated by unstable political alliances that formed the basis of ruling coalitions and hence overall turbulent and unstable political environment in the majority of the countries in the region. The generally unstable situation was also typical of Russia in the 1990s but was caused mainly by the policy of President Yeltsin who often encouraged conflicts and competition among political competitors and his subordinates in order to remove potential rivals and act as an arbitrator.

Distinctions between the two countries in the constitutional setting and peculiarities of their political regimes appear to have exerted a definite influence on the emergence by 2007 of the differing status of civil service reform implementation in Russia and the Czech Republic. The situation concerning continuation of the reform seems to have been more favourable in Russia where the President has been the major driving force of the reform and has relatively more powers in enforcing his policies (compared to the Czech executive authorities). In addition the Russian political parties as a rule play only symbolic role in the formation of the government and hence their interest in the right to "spoils" is limited.

The ruling coalitions in the post-communist Czech Republic used to be torn by internal contradictions and conflicts and on many occasions have demonstrated their clear interest in preserving the right to "spoils". Besides, administrative policy making in the Czech Republic is relatively more protracted (many decisions of a purely technical or organisational nature need to be pushed through the Parliament). Finally, after the Czech Republic joined the European Union in 2004, the major driving force of civil service reform – the European Commission - virtually ceased its pressure on the Czech authorities. All these factors considerably weakened the capacity of the Czech Republic to continue and sustain reform.

8.3.9 Economic Environment

In the early 1990s new political leadership in the majority of CEE and CIS countries adopted a strategy which focused above all on macroeconomic stabilization and microeconomic restructuring.

Development of a proper institutional and legal framework, in particular in the areas of public management, lagged behind considerably. The principle that "...the market economy is only 'free' within a public and legal set of enforceable rights and constraints" (Toonen, 1993:157) appears to have been largely disregarded. In fact neglect of this principle was one of the causes of serious recession in the Czech Republic in 1997-98 and financial default in Russia in 1998.

The implementation of economic reform strategies varied across countries in terms of speed and depth. The merits of fast reform versus gradual reform are still debated but as it turned out, almost all the governments in the region rushed ahead in rapid "big bang" style. The Russian Federation and the Czech Republic were no exception. Various studies indicate that there are certain and often striking similarities in the way privatisation was conducted in both Russia and the Czech Republic in terms of neglect of citizens' and public interests, disregard of the law and extent of favouritism. In the absence of proper legislation or its adequate enforcement bureaucrats often used their positions in ministries and other state bodies to advance their economic interests by establishing businesses in the very areas they were to regulate.

The World Bank data on transitional economies illustrate the trend in real gross domestic product (GDP) during transition. All transitional countries in CEE and CIS experienced output decline in the early 1990s: the extent of fall in GDP as well as the rate of recovery was for obvious reasons different in each country. It is noteworthy that the number of consecutive years of output decline between 1990 and 2000 for the Czech Republic and Russia were respectively 3 and 7. Cumulative output decline amounted to 12 for the Czech Republic and 40 for the Russian Federation.

Real GDP in 2000 was 99% of that in 1990 in the Czech Republic: correspondingly it was 64% in the Russian Federation. As one can see the slump (recession) in Russia was considerably greater in depth and duration (World Bank, 2002:5).

On the other hand, from 2001 till 2005 Russia has been demonstrating economic growth in the order of 5 - 6% per annum. This exceeds the respective figure of around 3% for the same period in the Czech Republic (World Bank, 2006).

It is interesting that comparison of government consumption as percentage of GDP in 1997 and 2004 shows approximately equal numbers 20.2% for the Czech Republic and 20.8% for the Russian Federation (ibid).

Summing up, throughout the 1990s the economic and budgetary situation in the Russian Federation and the Czech Republic did not differ substantially. The situation has notably changed in the early 2000s when the state budget of the Russian Federation started to receive extra revenues because of high rocketing oil prices on the world markets. This period practically coincided with the start of reform implementation in both countries.

8.3.10 Impact of Diffusion

Civil service reform agendas in both countries were largely formulated on the basis of information derived from the study of other countries' administrative experiences. Naturally modern trends in public management were taken into account as well.

Available information does not lead to a conclusion that a specific foreign model was used for preparation of the reform programme and strategy either in the Czech Republic or in Russia. Even so, it will be appropriate to consider the main sources of inspiration and experience that were used in designing civil service reform in each of the two countries.

Traditionally both Russia and the Czech Republic were following a continental European model of civil service. The only differences were that within the framework of this model Russia gave relatively more preferences to French experience, whereas the German system of public administration appears to be of relatively greater influence in the Czech Republic.

It should be noted (and we have shown this in the previous chapters) that elements of the so called New Public Management can be found in civil service reform programmes and in respective legislation in both countries.

In terms of specific impact on legislation, the first drafts of the Czech Civil service Act proceeded from the ideas of the Law on civil servants that existed in the Czech Republic during inter-war period. The latter was largely based on the Statutory rules adopted in the Austro-Hungarian Empire in 1914. In the opinion of Onisko the Czech territorial system of state administration was to a considerable extent copied from the Germanic states - Germany and Austria (Onisko, 2000 : 53).

It is known that in 1993 French experts assisted elaboration of the draft law on the Russian civil service (adopted in 1995). Later contributions to civil service reform came from various technical assistance and donor-aid projects and programmes. Starting from 2003 in Russia coordination of foreign assistance in the civil service reform area has been exercised by the Donor Secretariat on Civil Service Reform funded by DFID (UK).

Technical assistance from the EU to the administrative reforms in the Czech Republic came largely as part of PHARE programme. PHARE started operating in 1990 but initially only limited resources were devoted to public administration development. Similarly assistance from EU to Russia came within TACIS Programme which was launched in 1991. Until the early 2000s contribution to public administration development was relatively modest in terms of funding. Support was focused mainly on policy advice and civil servants' training. In recent years priority has been given to larger projects which are better coordinated, progressive from previous programmes and more adjusted to the specific needs of administrative and civil service reforms.

At the same time it should be emphasized that opportunities to benefit from foreign experience have not been always efficiently used. For instance, in the Czech Republic foreign contacts have been often limited to a very narrow group of specialists with the knowledge of respective languages. There was no national center for the study and research of public administration or for the analysis of policies and strategic studies in this area whether at government or university level. As a consequence, significant documents and studies, such as the Rotterdam document of the SIGMA programme, the EIPA study (Maastricht, 1997), World Bank country related research papers and many others were unfamiliar to the Czech authorities. Even the authors of various reform proposals often did not know them either (NTF 1998: 7-1-13).

Appropriateness of foreign experience for the Russian and Czech conditions has not been always properly assessed prior to its use. As a result some borrowed solutions did not work and often were even counterproductive just because they were designed for different environment. An example is decision of the Russian reformers to outsource some services rendered by state administration to non-governmental organisations or commercial entities. Under conditions of endemic corruption and absence of thorough preparation this move only facilitated abuse of authority and extended its scale.

Donor aid and technical assistance in the sphere of administrative and civil service reform was on average insufficiently coordinated and did not always take into account specific conditions of the recipient country. However it would be wrong to underestimate the impact of external assistance and diffusion. For example, support from international organisations allowed hundreds of civil servants from Russia and the Czech Republic to visit EU member countries and become acquainted with modern experience in the organisation and functioning of civil services and implementation of administrative reform.

Summing up the above we may conclude that:
a) attempts to reform the Russian and Czech civil services were based not so much on the appropriate and thorough analysis of the existing problems and shortcomings, but rather proceeded from the intention to conform to modern international trends in reforming public administrations;
b) there was no single source of inspiration for defining the reform dimensions and hence a rather *broad direction* approach was applied in both countries;
b) a certain measure of eclecticism and compromise in civil service reform proposals and the legal framework (both in the Russian Federation and the Czech Republic) can be explained in particular by weak coordination of reform efforts, impact of corporate and vested interests and general shortage of home based expertise.

8.4 Matching Civil Service Reforms

8.4.1 Reform Capacity

The analysis demonstrates that preparation of civil service reform in both countries took an extensive period – over a decade after collapse of the communist regimes in Central and Eastern Europe and the formation of the Czech Republic and the Russian Federation as sovereign states. This observation distinguishes Russia and the Czech Republic from the absolute majority of other CEE countries where civil service reform was launched much earlier.

We have established that the capacity for reform was overall low in both states. There are many indications of this low capacity and we considered them in detail in the respective chapters. The low reform capacity manifested itself above all in the following principal characteristics:

- absence of *legal status* of civil servants (until 2002 in the Czech Republic) or/and weak enforcement of the civil service related *legislation and normative acts* (in Russia since 1993 and Czech Republic since 2002 until present);

- strong elements of *patrimonial* bureaucracy prevailing in the Russian and Czech state administrations to the detriment of wide application of *meritocratic* principles in building up and managing modern civil services;
- domination of outdated informal rules, norms and traditions, in particular, widespread *clientelism and patronage;*
- excessive *politicization* of the two civil services;
- embedded *corruption* in Russia and the Czech Republic (more widespread in the former);
- absence of *political will* to launch civil service reform (was lacking throughout the 1990s in both countries);
- shortage of qualified *expertise and sufficient funding* (the latter valid for Russia till the early 2000);
- absence of committed reform stakeholders (e.g. mature civil society, influential business sector, strong political opposition, external pressure) that could force political leadership to adopt a pro-active stand towards civil service reform (in case of the Czech Republic no external pressure till the late 1990s);
- considerably weakened *competition* (often making it senseless) between major political parties and coalitions in the two countries;

The reform capacity was also influenced by unpredictability and uncertainty of the political and economic situation in both Russia and the Czech Republic (e.g. frequent change of governments, political and economic crisis, fragmented and conflicting legislation, etc).

On the positive side it should be emphasized that in both Russia and the Czech Republic there were small groups of experts who during the 1990s managed to prepare quality analysis of the status of the public administration of their countries and developed recommendations on how to improve the overall situation. Regretfully these analyses (Report of the National Training Foundation, Prague, 1998 and Concept of Administrative Reform, Moscow, 1998) were not properly reclaimed in the course of preparation of the civil service reforms.

During the late 1990s a number of attempts, though unsuccessful, were made in both countries by reform minded politicians and experts to convince their political leadership to launch civil service reform.

One of the substantial differences in terms of existing institutional environment and reform capacity between Russia and the Czech Republic was absence of a civil service in the Czech Republic from the legal point of view throughout all the 1990s. Other important differences include: a) in the Czech Republic the emergence in the late 1990s of the significant external factor of strong pressure from the European Commission to introduce civil service legislation and launch civil service reform resulting from the commitment of the Czech Republic to comply with strict requirements of EU membership; b) in the Russian Federation, significant changes in terms of existence of political will to launch civil service reform after V. Putin was elected President of the Russian Federation and committed himself to modernize Russian state administration; and c) also in the Russian Federation essential changes in some aspects of its economic situation in the early 2000s manifested in the high rate of economic growth and soaring state budget revenues mainly as a result of considerable increase of oil prices on the world markets. This development lessened budgetary constraints on the reform implementation.

Having compared the two cases we may conclude that the capacity for reform has been initially relatively higher in Russia than in the Czech Republic for a number of reasons:

a) contrary to the situation in the Czech Republic, where reform was driven mainly by external pressure from the European Commission, in Russia the reform's driving force was the President elected in 2000;
b) from the legal point of view the Russian civil service was established in 1993, much earlier than in 2002 in the Czech Republic;
c) starting from the early 2000s Russia was no longer experiencing tight budget constraints and could allocate considerably more funds for civil service development and reform.

In terms of the major cause (in broad sense) of initiation of civil service reforms in these two countries the answer is unambiguous. Profound transformation of their political and economic systems at the beginning of the 1990s made existing administrative structures obsolete and incapable of facing and countering new challenges and opened way for reform initiatives.

8.4.2 Reform Implementation

As far as implementation of reform is concerned the two states have demonstrated a top down approach to initiation and carrying out of civil service reform. In view of the atmosphere of confidentiality surrounding reform preparation and evident shortage of information and PR strategy to raise public awareness about the reform, not only citizens, but also the majority of civil servants were largely unaware of reform objectives, agenda, time table, benefits, progress etc.

Civil service reform agendas in the two countries were formulated mainly on the basis of information derived from the study of modern trends in administrative reforms and other countries' experiences in this field. Therefore it comes as no surprise that the contents of reform programmes in many ways were similar. They comprised measures directed mainly at raising the efficiency of state administration and creating in the civil service a culture, focused on performance and outcomes and on service to citizens;

More specifically the programmes of civil service reform in Russia and the Czech Republic (complemented by elements of administrative reform where necessary) comprised the following major tasks to be achieved: a) carrying out functional analysis; b) re-organizing the Government; c) improving coordination and communication; d) defining standards and regulations; e) introducing modern management technologies, above all performance management; f) designing and enforcing civil service law; g) promoting e-government practices; k) enhancing training of civil servants; l) developing quality management; m) improving financial management; and n) developing anti-corruption mechanisms.

In both countries civil service reform was regarded as complementing or being part of a wider administrative reform. The specificity of the Czech case was that for quite a long time public administration and civil service reforms were strongly associated with the reform of the territorial and regional government.

The approach to reform implementation in the Czech Republic was based on the application of principles of project based management. In the Russian Federation implementation was driven through the so called federal targeted programmes. The difference between these approaches is not essential. Both instruments are known for strong result orientation if properly and consistently applied. Federal targeted programmes appear to be more centralized, whereas project based management allows for greater de-centralization.

However, practical implementation of these instruments both in Russia and the Czech Republic turned out to be problematic. Cases of better performance and achievement of expected results appear to be rather exceptional so far and confined to individual organizational units but very rarely to larger institutions (e.g. ministries) not speaking about entire civil services.

A major paradox of civil service reforms in Russia and the Czech Republic is that, although practical introduction and application of performance and results oriented management were declared the principal objectives of administrative transformation, in managing the reforms themselves these objectives were *de facto* largely ignored.

Implementation of civil service reform took place mainly in central administration in the Czech Republic and in both federal and regional government in the Russian Federation. With regard to reform implementation tools, exclusive preference in both countries was given to legislation. Other tools such as training and best practice diffusion were also applied though not always on a wider scale. To a considerably less extent were used fiscal incentive instruments, networking and coalition building. Limitations in the use of implementation tools reduced the potential of reformers in both countries to reach reform objectives.

In Russia and the Czech Republic the reform strategy implied launching a series of pilot projects before full-scale implementation was attempted.

In differentiating the civil service reforms according to dominant tradition – *governance* or *management* (Ingraham) - we need to take into account not only the contents of the reform programmes but also distinguish declared objectives and real achievements in practice. Since the development of state administrations in both Russia and the Czech Republic was rooted in the continental European tradition of state bureaucracy one would expect that for the reformers in both countries the *governance tradition* will be the main source of inspiration.

This tradition in fact was paid tribute to in the reform programme declarations. But overall the reform agenda appears to be mixed containing priorities adhering to both *governance* (civil service is primarily a mechanism for effective governability and political legitimacy) and *management* (attention above all to efficiency and economic rationality) traditions. Furthermore, in practical implementation of reform in both countries distinct priorities were given to the measures directed at strengthening aspects of efficiency of civil services. Issues of legitimacy, accountability and transparency in their practical enforcement turned out to be quite “unpopular” among bureaucratic officialdom in both Russia and the Czech Republic.

In connection with this, classification of the models of civil service system that are being reinforced within the framework of reform implementation in Russia and the Czech Republic encounters certain difficulties. Within the models framework advocated by Pollitt and Bouckaert this “mixed” combination of reform initiatives typical of both reviewed cases tends to place them half way between New Public Management (NPM) and Neo Weberian State (NWS) concepts.

Categorization of civil service reforms in line with the criteria defined in the Protocol for comparative studies of national civil service systems leads to the conclusion that they were partly *structural* and partly *procedural/technical.*

The extent of intensity of reform measures and pace of reform progress appear to have varied. The restructuring of the Russian Government (as part of administrative reform) was quick and most radical at least for the last 40-50 years of the Russian history. Implementation of reform of the Czech civil service was advancing much slower. Quite demonstrative in this respect was the decision of the Czech authorities to suspend several times full enforcement of the Civil service Act. The time lag between the adoption of the Act in 2002 and numerous postponements of its full enforcement with the latest one extending the law's "invalidity" till 2009 is seven years which appears unprecedented by any standards. After an initial stage of quick advance the pace of the reform in Russia has also slowed down.

Management of civil service reforms necessitated appointment of the highest authority and establishment of special structures responsible for it. In the Czech Republic this was the Prime Minister and Office of the Government strengthened at the operational level by the Civil Service Directorate and Steering Committee to oversee and supervise reform projects. In the Russian Federation this was an ad hoc Commission on Reform of Governance headed first by the Prime Minister and then by the Head of the Presidential Administration. The Commission was working in close interface with the Presidential Administration and several expert groups. Reform implementation in the two countries lacked sound management mechanisms at the operational level, as well as proper and timely monitoring and progress assessment procedures, though the latter was overall better in the Czech case. Bureaucratic resistance to change (although generally anticipated) was not properly taken into account, even less so was effectively dealt with in both countries.

The Russian authorities were quick in adopting legislation but many of the innovative provisions have remained so far on paper because of weak law enforcement. Reform results in the Czech Republic appear to have been confined so far mainly to pilot experiments.

A peculiarity of the Czech case is that the reform strategy and programme were developed and approved long after the Civil service Act was adopted. In other words co-relation and congruence between the legislation and practical implementation work were difficult to achieve. The situation was further compounded by numerous postponements of the full enforcement of the Civil service Act.

8.4.3 Reform Outputs

Further comparison of the two reforms leads us to the following reflections and inferences. Throughout the entire analysis we have made regularly the point that the task faced by the reformers in Russia and the Czech Republic was not in improving, reorganizing or modernizing their civil service, but in essence in its *re-founding*. One of the conclusions reached in this research is that this need has not been fully acknowledged by the reform managers in Russia and the Czech Republic and even less so taken into account in practical activities.

Preparation of the Czech Republic for EU membership provided its representatives with many opportunities for intensive dialogue and cooperation with the European Commission and its officials and experts. Commitments taken by the Czech authorities to meet criteria of EU membership also in the sphere of administrative capacity and quality of civil service were demanding and stimulated modernisation of the Czech state administration in a number of fields. Russia could not benefit equally from these opportunities but was open to various programmes and projects of international assistance.

Since in neither of the countries civil service reforms were officially announced to be finished nor the time table of reform implementation expired by the time of finalizing this research, we can only speak about preliminary results of reform implementation. Overall these appear to be modest.

Obviously the fact that civil service reforms were eventually launched in Russia and the Czech Republic can be viewed as an achievement in itself. Undoubtedly it raised awareness of the expert community and civil servants in each country of the existing problems and ways of resolving them.

Overall, introduction of the principles of meritocratic, performance oriented, uniform, integrated, accountable and corruption immune civil service so far has not been sufficiently consistent and result oriented in either country. Therefore many expected outcomes turned to be less tangible or intangible or failed to materialize at all.

Among tangible reform results **in Russia** so far are – adoption of new legislation on civil service and related normative acts (e.g. Code of Ethics), increase in civil servants remuneration package (but with rather weak link to performance), large scale restructuring of the government (as part of administrative reform), wider scale of training of civil servants, and greater transparency achieved through the web-sites created by virtually all federal ministries and agencies. A number of processes were initiated – review and development of qualitatively new job descriptions, functional analysis of the ministries and structural units, development and approval of standards of services rendered to citizens and legal entities, introduction of the principles of performance based budgeting, etc. However, it is too early to speak about the success of these initiatives.

One of the peculiarities of the civil service reform in Russia is that implementation of civil service related laws requires preparation and adoption of the so called secondary legislation – Presidential decrees, Government enactments, Federal Ministries' orders - which specifies the mechanism of enforcement of the provisions contained in the Laws. So far this process has been developing at a much slower than expected rate.

Simultaneously a number of experimental pilot activities were carried out on the development and introduction of performance related pay, performance budgeting and performance management. In 2006 a Presidential decree allowed several government agencies and ministries (as an experiment) to retain the overall pay bill of those employees who have left public office and to use the savings for motivating quality work of the remaining personnel. There has been also increase in state budget funds allocated for civil servants training. In coordination with administrative reform certain practical measures were taken to reduce administrative barriers for businesses. This include: reduction of the powers of licensing and controlling organs by introducing, for example, the right to open business in five days period following submission of a fixed number of legal documents; considerable reduction of control and inspection exercises by introducing fixed frequency and scheduling; reducing the list of business areas that need to be licensed from 214 to 120; and some others (Otechestvennyie zapiski 2004:107).

Some important measures envisaged in the new legislation and the entire reform programme have not found their way to practical implementation yet. For example, commissions on conflict of interest were created but no single case has been reviewed in course of at least two years of their existence. Fixed term contracts with higher remuneration closely linked to performance as envisaged by the Law, have not been applied in practice yet. Similarly, creation of a body in charge of managing the entire civil service was still pending by late 2008. Some draft laws e.g. "On Access to Information About Government Activities" or "On Fighting Corruption" were prepared and even discussed in the Parliamentary commissions but later repeatedly pushed from one desk to another for several years without visible progress to final adoption.

The precise extent of reform advance is difficult to judge as the reform programme (at least the variant accessible to the public) did not contain specific deadlines for implementing each of the defined tasks. Many tasks were not sufficiently detailed. Deadlines could be found in the reform action plans adopted by some of the federal ministries. However, there does not seem to exist any document comprising results of regular monitoring of the progress made in reform implementation across the entire federal government.

Up to date major outputs of civil service reform **in the Czech Republic** (called reform of the central state administration in Prague) incorporate: adoption of the Civil service Act; establishment of the General Directorate for Civil Service (so far understaffed and lacking support to its mission) which was supposed to be responsible in future for management of the entire civil service; creation of the Institute of Public Administration; increase in the number of civil servants who has been trained; and wider use of IT technologies.

Looking at the reform from a wider perspective one should add here also establishment of Administrative Judiciary, adoption of the Act on Access to Information, approval of the Code of Ethics for government officials.

So far the reform process has focused chiefly on the improvement of performance of central state administration, regulatory reform and strengthening financial control in central state administration. The tangible results of the projects implemented within the framework of the reform programme are mainly methodological instructions and manuals on how to introduce various types of innovations and improvements e.g. CAF quality assurance (Vidlakova, 2006). Information about wide dissemination of the project outputs and deliverables across entire government is not available. Subsequent developments, namely actual suspension of the reform by 2007 provide sound ground for argument that this dissemination has hardly taken place on any noticeable scale.

8.4.4 Two Civil Service Reforms in Mirror Comparison

The results of comparison of current experiences of civil service reforms in Russia and the Czech Republic are summarized in Table 8.2 and are reflected in certain conclusions that follow the table:

Table 8.2 Comparison of Civil Service Reforms in the Russian Federation and the Czech Republic

Characteristics of reform	Russian Federation	Czech Republic
Type and model of reform	Structural and procedural -technical Elements of governance and management models, the latter generally dominating in reform practical activities	Structural and procedural -technical Elements of governance and management models, the latter prevailing in the reform strategy and practice
Reform objectives	Creation of an integrated civil service system; Increase of the efficiency of activities of civil service; Formation of highly skilful and competent personnel; Creation of modern system of civil service management. The reform top priority - *improving **performance** of civil service, optimizing civil service maintenance costs and developing resources required for civil service activities.*	Legal establishment of civil service as such; De-politicization of civil service; Enhancing stability and professionalism; Raising efficiency of civil servants' activities; Introducing proper management of civil service. The reform top priority – *enhancing **efficiency and quality** of services rendered by civil servants to the state, business and citizens*
Approach to reform	Top down	Top down
Managing authority	Inter-Agency Commission on Improvement of State Governance headed at first by the Prime Minister, later by the Head of the Presidential Administration. Coordination – by the Department of Civil Service in the Presidential administration. Operational management at the level of designated ministries.	Overall Prime Minister. Coordination and supervision by the Head of the Office of the Government who is supported by the Department of the Regulatory Reform and Central State Administration Reform. Operational management - by the Reform steering committee (management team).
Reform concept and programme	Concept (2001) and Programme (2002) approved by the RF President, the Programme extended in 2005 for another two years. Draft Programme for the years 2008-2013 was submitted for discussion.	Concept (2001) and Programme (2004) approved by the government, later the programme underwent several modifications.

Extent of reform	Generally across all levels of government including federal and regional levels	Encompassing only central government structural units
Civil service legislation	The first law approved in 1995 but a number of its provisions were enforced only partly. New laws on civil service were adopted respectively in May 2003 and July 2004.	Civil service law (Act) – adopted in April 2002, few provisions valid since 2004, full enforcement several times put off, latest deadline fixed for 2009. Finally cancelled in August 2007.
Reform main instruments	Federal targeted programmes	Project management method
Impl-n tools	Definite priority given to legislation, less attention to best practice diffusion and fiscal incentive instruments, networking and coalition building have been practically disregarded	Prevailing accent on legislation, less attention to best practice diffusion and training; fiscal incentive schemes, networking and coalition building have been practically disregarded
Reform status	Reform Programme extended till 2008 by a Presidential decree. Draft Programme of civil service reform and development for the period 2008-2013 submitted for consideration.	In October 2006 reform management was formally passed back to the Ministry of Interior, reform scope and contents have been considerably reduced and modified, in essence reform implementation was put on hold.
Original time frame of reform impl-n	From 2003 till 2005 and further to 2008	From 2004 until 2010
Impact of diffusion	Overall considerable, at the level of design and implementation	Considerable, at the level of planning and implementation
Intensity of reform	Rather high at the start but later slowed down	Never high, rather slow advance with small steps
"Active" and "passive" internal and external stakeholders	"Active" – Mr. Putin, President of the Russian Federation; "passive" - business community, civil society.	"Active" – Mr. Havel, President of the Czech Republic, European Commission; "passive" – business community, civil society.
Achievements to date	New legislation, reorganization of the government, higher compensation package for civil servants, methodological guidelines for the introduction of new approaches and innovations, experiments on	Legal status of civil servants established, Institute of Public Administration set up, methodological guidelines for the introduction of various innovations developed, IT technologies used on a wider scale,

	introducing performance based pay and standards of service, more funds allocated for civil servants' training	expanded programmes of civil servants' retraining.
Prospects for reform continuous advance and reaching its principal objectives	In the medium term overall positive, in the long term risk of loosing political support after Presidential elections in 2008 and in case of economic recession	Rather dim and uncertain, since in late 2006 the reform appears to have been removed from the list of Government priorities and seriously emasculated. Further undermined by abolition of Civil service Act in 2007.
Unforeseen outcomes and side effects	Considerable increase in budgetary spending on state bureaucracy and in the number of civil servants at the federal and regional level; Expansion of privileges and bonuses granted to middle and high level civil servants; Emasculation of some innovative reform proposals producing often opposite than expected effect;	Increasing concern of the politicians that they may lose control over the state bureaucracy led to actual suspension of most reform activities; Considerable reduction of the contents and scope of reform. Eventual dissolution of reform's original management organs and structures. Abolition of Civil service Act in 2007.

On the positive side it should be stressed that after a long period of uncertainty and hesitation civil service reforms in both Russia and the Czech Republic were actually launched and achieved certain results. In the process they drew attention to the existing in public administration problems, contributed to widening of cooperation in this field with international organizations and donor community, raised the number of qualified local experts and increased the number of civil servants who have been trained at home and abroad.

A number of pilots and experiments were initiated, some of them successfully implemented.

Both reforms are distinguished for their *focus on enhancing efficiency and performance* which have been declared the principal objectives and appear to have been pursued in practice (at least at the start).

At the same time careful study of the processes of reforms' preparation and implementation reveals that up to date there has been *no real ownership of reform* in both countries. Reforms were viewed as something imposed externally, either by top authority (the President in the Russian case) or by foreign community (European Commission in the Czech case). This lack of ownership was reflected also in the actual absence of influential reform stakeholders. For example, neither business community nor civil society appear to have played any prominent role in voicing their support to reform and determining its pace and contents. Lack of ownership in both countries was also manifested in the sheer shortage of leaders who could move beyond traditional perceptions of management and display resoluteness and consistency in driving innovative approaches.

The reforms did not succeed so far in gaining *broad support* from civil servants, citizens and political parties. At least partly this can be explained by the fact that in both countries programmes of civil service reform and respective legislation on civil service were prepared in the atmosphere of strict confidentiality. In practical terms reform strategy lacked such important elements as raising awareness, explaining benefits, rewarding active participation etc.

Both reforms were not sufficiently civil service staff focused and were relying mainly on changes driven externally. In particular, the announced development of modern HR management systems and introduction of performance based incentives remained until present at the stage of elaboration of methodological support and narrow scale experiments. Most importantly the *reforms failed so far to create motivation* among the majority of civil servants and politicians *to support new policies and changes.*

The weakest point of reform management and strategic approach in both cases turned out to be excessive and even exclusive *reliance on legislation as the main instrument of reform implementation.* Passing laws achieves little if there is no effective mechanism for its enforcement and no programme to reach strategic objectives. In fact, the Russian and Czech cases both provide good examples of how law making was disconnected from its implementation. It appears that authors of numerous laws and normative acts were convinced that their enforcement would take place automatically. Little attention has been paid to developing specific mechanisms and methodologies for implementing adopted legislation and reform programme. The situation appears to be even worse when enforcement of adopted legislation is permanently put off as has been the case in the Czech Republic.

Comparing further various aspects of reform preparation and implementation in the two countries we may conclude that the Czech civil service reform programme of 2004 (though heavily emasculated by 2007) appears to have been from the very beginning more result oriented. The programme of reform of the Russian civil service was supposed to be broken down into programmes in each ministry, agency, and regional administration. In each domain particular responsibility and deadlines for meeting specific objectives were as a rule defined. In contrast the Czech reform programme from the very outset comprised specific outputs and deadlines (applicable for the entire central administration), references to responsible persons, regular reporting and monitoring procedures, and actions to be taken should implementation be delayed or encounter serious problem.

Therefore it appears at first sight illogical that by late 2006 the civil service reform in the Czech Republic had been virtually scale downed and put on hold after approximately just two years of implementation. No other conclusion, except suspension of reform, can be drawn after overall management of the reform was passed over back to the Ministry of Interior, existing reform management structures disbanded and the original contents of the programme considerably reduced. Out of the original list of 17 reform projects 5-6 were announced terminated (but no account of their impact released) and only three projects (considerably modified) appeared to be in process of implementation at early 2007. The new Institute of Public Administration initially created within the structure of the Office of the Government was also transferred to the Ministry of Interior.

The Civil Service Directorate (key management structure for the entire civil service created in June 2002) by 2006 seems to have lost any support for its mission with its staff reduced from 65 to 22 persons (Landova, 2006:60).

This is even more bewildering since before the reform was launched in 2004 its preparation had been managed by the Ministry of Interior. The continuation of this practice was declared by the Czech authorities to be unacceptable because any single ministry is not in an appropriate position to lead reform efforts targeted at the entire government structure.

Taking into account the repeated postponement of the enforcement of the Civil service Act one cannot help concluding that by late 2006 remnants of any formal commitment of the country's leadership to civil service reform in the Czech Republic had disappeared. This conclusion was further reaffirmed in August 2007 when the Czech Government decided to abolish the Civil service Act all together.

Therefore, one of the greatest variances between the civil service reforms in the Russian Federation and the Czech Republic relates to their status by late 2006 and early 2007. Contrary to the situation in Russia the civil service reform in the Czech Republic had actually stalled by early 2007.

Among other significant distinctions between the two reforms the following should be mentioned:

a) contrary to the Czech civil service which formally and legally had not existed before 2002, the Russian civil service was established in 1993;
b) in contrast to the Russian Federation, public administration reform in the Czech Republic was viewed until late 1990s exclusively as a reform of territorial government;
c) our analysis revealed that by the early 2000s overall capacity for civil service reform was relatively greater in the Russian Federation;
d) the major driving force of the reform in Russia was the RF President elected in 2000 whereas in the Czech Republic it turned out to be external factor - the European Commission;
e) civil service reform in the Russian Federation extended also to the regional level whereas in the Czech Republic it has been limited so far to the central state administration only;
f) irrespective of the eventual results of enforcement of civil service related legislation, the original objective of the Russian Laws (2003 and 2004) was to develop further a career based civil service, whereas the Czech Civil service Act was designed to introduce a sort of mixed career and position based civil service system;
g) there appears to be a marked difference in the pace of reform advance at least during its start up phase - slow and cautious, bordering on "no movement at all" approach to reform of the central state administration in the Czech Republic (2004 – 2005) and rather "big bang", "shock like" approach in restructuring the Russian Government in 2004. Later this approach turned into relatively more gradual intervention as far as implementation of the rest of the Russian civil service and administrative reform programmes is concerned;
j) contrary to the situation in the Czech Republic, starting from the early 2000s there has been notable relaxation of fiscal pressures on the civil service reform in Russia resulting mainly from the considerable increase of oil prices on the world markets, contributing to high rate of economic growth and soaring state budget revenues and hence better opportunities of funding reform activities.

A number of additional observations should be made with regard to reform implementation in Russia and the Czech Republic. It appears that implementation of reforms have contributed to the appearance of a number of distinct *side effects and unforeseen results*. Reorganization of the Russian Government in 2004 virtually paralyzed the activities of the executive branch for at least half a year. It also resulted in exodus of a notable number of professionals from the civil service. A number of other developments cannot be directly attributed to the impact of reform but they were against its spirit and purpose; the reformers could not prevent the developments happening. The glaring example in Russia is the considerable increase in budgetary spending on state bureaucracy and in the number of civil servants, particularly at the regional level, that happened in the years 2002–2006. Another case relates to expansion of privileges and bonuses especially for top level civil servants. For instance, in 2006 Russian civil servants occupying positions of Deputy Head of Department and upwards were granted the right to business class travel by air and first class travel by railway.

As far as the Czech Republic is concerned the major "unforeseen" result (from the point of view of the reform logic but of course not quite unexpected from the point of view of many experts studying the Czech political scene and administrative traditions) is the suspension of civil service reform before any principal outputs have been produced.

Also typical of both reforms was the fact that issues of *fighting corruption* turned out to be at the periphery of attention of reform managers. There has been much rhetoric about the need to eradicate fraud and conflict of interest but very little (if anything) has been done so far in terms of achieving practical tangible results.

Table 8.3 Priorities of Civil Service (administrative) Reform Programmes in Russia and the Czech Republic and Extent of Their Implementation (by September 2007)

Some of the priorities of the reform programmes	Up to date results Russian Federation	Up to date results Czech Republic
Implementation of functional analysis (identifications of missions, functions and analysis of processes) in the government agencies	Identification of overlapping and excessive functions largely carried out, but liquidation of these redundant functions turned to be extremely difficult. Auditing of processes is in pilot phase	Identification of missions and functions was done in 2005 but analysis of processes (project C2 - process audits) failed, since it was carried out only in one ministry and later declared no longer compulsory
Defining standards of services rendered to citizens and legal entities	Limited number of standards introduced in a few regions, overall at the experimental stage	Pilot experiments in selected agencies
Reorganization of central government	Carried out in 2004	Cancelled (project A4)
Depoliticizing civil services	No tangible progress apart from generally more clear distinction between political and administrative posts	Largely depended on full enforcement of the Civil service Act; following cancellation of the Act the prospects are quite obscure
Introduction of merit based principles of service (open and fair competition, performance based pay,	Stipulated by the Law but cases of open and fair competition and pay directly linked to performance are	Open and fair competition exercised in a number of government agencies, wide application of other

modern job descriptions)	still rare, whereas modern job descriptions are more widely used	principles is questioned following abolition of the Civil service Act
Adoption of civil service laws and other relevant legislation	New Laws adopted in 2003 and 2004, more than 20 Presidential decrees on civil service issues signed since the start of reform	Civil service Act adopted in 2002, more than a dozen amendments since then, but its full enforcement postponed several times, until the Act was abolished in 2007
Remuneration of civil servants	Since the start of reform was regularly increased at the rate higher than average for other groups of employees, particularly large for top level civil servants	More or less regularly adjusted to the level of inflation
Establishment of central body/unit for management of civil service	Still pending although stipulated by the Law	General Directorate of Civil Service was established in the Office of the Government in 2002 but failed to function properly because of lack of support to its mission and still pending appointment of its Head.

The analysis of the recently adopted *laws on civil service* in the Russian Federation (2003 and 2004) and the Czech Republic (2002) done in the previous chapters shows that they have much in common, though the Czech Civil service Act appears to be considerably larger in size and generally more complicated in perception.

The Czech law has 254 articles and three annexes. But only 15 articles and one annex was enforced upon promulgation (Vidlakova, 2006: 2). By the year 2006, thus even before full enforcement of the Law it had been amended seven times! (ibid). By comparison the draft law from 1993 contained almost 400 pages (275 articles) plus attachments (Cebisova, 1996:136). The respective Russian laws were more compact, for example the one adopted in 2004 contained 74 articles and 65 pages.

Comparison of the new laws reveals that they:

a) establish the fundamental legal framework for civil service activities;

b) comprise detailed provisions with respect to definition of civil servants, their rights and obligations, restrictions and benefits, conflict of interest, admission requirements, discipline and penalties, job classification and duties, ranks and grades, training and professional development, performance evaluation, compensation and service termination; but are much less clear and specific with regard to civil servants' responsibilities and external accountability, introduction of competitive practices, exercising transparency and openness;

c) are largely oriented towards meeting corporate interests of civil servants and ambitions of political elite to the detriment of their commitment to the state and citizens;
d) the Russian Law creates the basis for the development of a career based civil service whereas the Czech Act introduces a sort of mixed – career and position (at least on paper) civil service system.

The greatest difficulty concerning new civil service legislation in both countries rests in the problems of their effective enforcement.

8.4.5 Key Variables and Determining Factors

At the start of our analysis (Chapter II) we asserted that our selected variables may be directly or indirectly linked to the character and essence of the political regime and development of political institutions, resoluteness of political leadership to launch reform and be a driving force for changes, type of relations between the policy makers and top civil servants, extent of domination of interest groups, prevailing behavioral patterns, traditions, informal norms and rules, maturity of civil society, extent of internal and external pressure, and available resources. Having finalized our research we can make the following conclusions.

Our study allowed to confirm the impact of these variables (though to a varying extent) on the development and reform of administrative institutions in Russia and the Czech Republic.

In pursuit of answers to the research questions we have found out that clientelistic and patronage networks (widespread in the former communist countries) survived political upheavals and have remained the base of relations across state administrative institutions in the Russian Federation and the Czech Republic. These networks are "reinforced" by high degree of politicisation of the Russian and the Czech civil services. Political parties and leaders view state administration as an instrument of their personal power. In most instances members of such networks tend to maintain mutually beneficial status quo, support broad opportunities for discretion and resist any change.

These features, tendencies and behaviour are rather common and can be found in other countries. However they "flourish" only under specific conditions. The best and well known "remedy" to treat such "deviations" is to foster political competition and accountability. The latter is effective only where there exist transparent and fair rules of competition and genuine responsibility. One of the conclusions reached in this study is that the rules of political competition and accountability, though generally proclaimed in the constitutions and other laws, have not been appropriately enforced in Russia and Czechia. This appears to be one of the key factors that "contributed" to delays in launching civil service reforms, virtually uncontrolled growth of the number of civil servants and achievement so far of rather modest results in the reform implementation in both Russia and the Czech Republic.

The prospects for successful reform depend to a large degree on the availability of political will to effectively tackle the problems of civil service development. However, even of greater significance is the appropriate political and institutional framework that ensures timely adoption of required polices and decisions even in cases when such political will is missing. Also critical for success of reforms is the existence of a mature civil society and independent media.

The analysis revealed both similarities and differences between the Russian Federation and the Czech Republic in the ways reforms were prepared, launched and implemented.

Similarities in reform preparation and implementation can be in particular attributed to the largely common roots of administrative culture and mentality which can be traced to the heritage of monarchist (Austro-Hungarian and Russian Empires) and recent communist regimes.

The causes of major differences between the two countries in the civil service reform processes and current outcomes, above all suspension by 2007 of the reform in the Czech Republic and continuation of the reform in the Russian Federation, are largely rooted in the distinctions in the existing institutional configurations of the two states.

Our study also confirms that the role of leadership is crucial for the success of civil service reform. It is even more important when reforms follow a top down scenario. Therefore preferences and personal qualities of politicians (who act under specific conditions and whose discretional power is often limited) do matter. Following our analysis we may conclude that the largely uncompetitive political environment that existed in Russia and the Czech Republic throughout most of the 1990s reduced to minimum opportunities for correcting the policies of the political leadership (to be more precise absence of such policies) regarding reform of civil services.

There is sufficient evidence to claim that these policies were eventually "corrected" by: a) the pressure from the European Commission in case of the Czech Republic; and b) the election of the new President in 2000 in case of the Russian Federation.

Our study has also reaffirmed our assertion that no matter how radical and innovative declarations of reforms in Russia and the Czech Republic could be (under the influence of either internal or external factors), the extent of their implementation may substantially deviate from proclaimed intentions. In particular we realized that the political and bureaucratic elites in the two countries had largely simulated and distorted the essence of civil service reform that could limit their discretion. There has been strong tendency to confine civil service reform to purely legal and organizational (technical) issues and to make adoption of civil service law the starting and finishing point of the reform.

8.5 Conclusions

The outcomes of the theoretical and methodological analysis (Chapter II) and the inferences made in subsequent country specific chapters (Chapters III-VI) have allowed us to distinguish factors and forces (variables) that have exerted serious influence on the processes of civil service reform preparation and implementation in the Russian Federation and the Czech Republic.

Within the identified set of variables certain factors appear to have played a decisive role in defining the actual direction and pace of reforms as well as determining the results achieved so far. These are excessive politicization of the respective civil services, wide spread patronage networks, and the administrative culture and leadership factors. A distinct feature of these key variables is that their origin is heavily influenced by historical legacy and path dependence.

By matching the two civil service reforms we were able to define commonalities and distinctions in their contents, driving forces, rate of advance and outcomes. Analysis of the impact of the above mentioned set of variables on the processes of civil service reform preparation and implementation in Russia and the Czech Republic and comparison of this impact in these countries allowed us to trace the roots and to identify basic causes of the reform related developments and constraints.

In brief, the cumulative effect of the long term impact of the identified variables resulted in the overall weakened reform capacity in Russia and the Czech Republic throughout the 1990s though this capacity was relatively higher in Russia. This impact led to a situation where clientelistic networks (widespread in the former communist countries) survived political upheavals and remained the basis of relations across state administrative institutions in the Russian Federation and the Czech Republic. These networks were "reinforced" by high degree of politicisation of the Russian and the Czech civil services. In most instances members of such network tended to maintain a mutually beneficial status quo and resist any change. Political parties and its leaders continued to view state administration as an instrument of their personalistic power. The existing administrative culture was hostile to the principles of merit based career advancement and compensation giving preferences to the principles based on length of service and loyalty to the superior management.

The impact of the same variables determined the emergence of many weaknesses and constraints in the processes of implementing civil service reforms in Russia and the Czech Republic and hence appearance of many similarities in these processes and their results. There are also differences, some of them essential. For instance, there are differences in the reform driving forces and in the status of reform implementation by 2007. Our analysis revealed that these differences largely stem from the distinctions in the existing institutional configurations of the two states.

As an illustration one can consider the peculiarities of the constitutional framework and political regimes in Russia and the Czech Republic. On the one side, the Russian President, the major driving force of the reform, with relatively more powers (compared to the Czech executive authorities) in enforcing his policies, and political parties in Russia with a largely symbolic role in the formation of the government and hence making virtually no claim for the right to "spoils". On the other side, the ruling coalitions in the Czech Republic, frequently torn by internal contradictions and conflicts, clearly interested in preserving the right to "spoils", often constrained in policy implementation by the need to push even minor organisational decisions through the Parliament, and finally no longer heavily pressed by the European Commission to keep civil service reform going.

These systemic and situation specific peculiarities of the institutional configurations in the two countries largely explain the difference in the status of the reform by 2008, i.e. continuation of civil service reform in Russia and its suspension in the Czech Republic.

CHAPTER IX
FINDINGS AND CONCLUSIONS

9.1 Introduction

In the theoretical and methodological part of our research we have demonstrated that civil service reform is a complex process which cannot be studied in isolation from its environment and existing institutional configurations. Therefore we considered it most appropriate to apply *institutional analysis* in the study of the development of civil service systems which constitutes the core of this research. It implies that study of existing rules, norms, values and authority relationships which exert significant influence on civil servants' behavior and persist in time is a necessary precondition for proper understanding of civil service systems, their reform and evolution over time.

It also implies that various institutions define and limit choices available for policy-makers. At the same time the role of individual and collective human action should not be underestimated.

In the early 1990s in the atmosphere of overall fascination with market reforms in Central and Eastern Europe there was general lack of apprehension of the importance of creating professional, efficient and accountable state administration. This initial underestimation extended also to donor agencies and technical assistance providers. The objective of building up a modern and effective system of governance, complemented by checks and balances, control and supervision, ensuring stability and performance orientation of the civil service was largely neglected.

Until the early 2000s the highest authorities in Russia and the Czech Republic effectively ignored civil service reform in terms of its practical implementation. During the decade that followed the demise of communism practically all political, economic, social and other institutions, except state administration, have undergone profound changes in both Russia (see e.g. Aarrevaara, 1999; Kotchegura, 1999) and the Czech Republic (see e.g. NTF 1998; Potucek, 1999). It is true that some structural alterations took place for example in the Russian and Czech governments. However, as we have shown in the previous chapters, the rigid bureaucratic spirit, outdated administrative practices, patronage based networks remained largely untouched.

In essence, a kind of contradiction has gradually built up in both Russia and the Czech Republic. The political and economic systems of both countries have drastically changed but the way these countries had been managed administratively was largely retained.

The civil service reforms that were finally launched in Russia and the Czech Republic in the early 2000s are distinguished by both differences and similarities. Identification and comparison of the determining factors that affected reform preparation and implementation in the two countries led us to the following conclusions.

9.2 Current Reform Results and Explanatory Factors

Our analysis demonstrated that various forces, factors, tensions and constraints have influenced the process of civil service reform initiation and implementation in Russia and the Czech Republic.

Throughout the 1990s the Russian Federation and the Czech Republic were demonstrating overall low capacity for civil service reform across all of three major parameters:

a) institutional space (framework) - formal and informal institutions (combination/set of rules, norms, traditions and values) existing at present and historically;
b) collective or individual human action (willingness and readiness of the key agents – policy makers and stakeholders to undertake action in the desired direction);
c) resources available (funds and expertise).

The low reform capacity was to a considerable extent predetermined by the historical development and legacy, national and administrative traditions and "path dependency" factors. Widespread patronage and clientelistic networks, excessive politicization of civil services, clear deficit of meritocratic principles, prevailing culture and mentality rooted in the communist and pre-communist past, insufficient external accountability – all this reduced considerably the internal potential for civil service reform in both countries.

In Russia and Czechia the development of new political and state institutions in the late 1980s to early 1990s was happening not in a "bottom up" mode as a result of systematic and consistent struggle of large groups of society for their rights and freedoms (e.g. *Solidarity* in Poland), but mostly in a "top down" fashion, through "imposition" or "implantation" of these institutions by the post-communist elites. To a considerable extent this situation explains the slow development of civil societies in these countries and wide opportunities for discretion of their political elites.

The inherited and deeply rooted habit of the Russian and Czech political elites *to treat the state institutions and civil service as the instrument of their personal power* appears to have survived the political turbulence of the 1990s and to have influenced the development of the civil services in these countries.

The situation was further compounded by lack of political will (at least until the early 2000s) of the ruling elites to implement civil service reforms and virtual absence of committed reform stakeholders. Finally, throughout the 1990s Russia and the Czech Republic experienced severe budgetary constraints and shortage of home based expertise.

Having compared the two cases we may conclude that the capacity for civil service reform has been initially relatively higher in Russia than in the Czech Republic for a number of reasons:

a) contrary to the situation in the Czech Republic, where reform was driven mainly by the external factor of pressure from the European Commission, in Russia the reform's driving force was the RF President elected in 2000;
b) from the legal point of view the Russian civil service was established much earlier - in 1993, whereas the Czech civil service only in 2002;

c) from the early 2000s Russia was no longer experiencing tight budget constraints and could allocate considerably more funds for civil service development and reform.

Civil service reforms in Russia and the Czech Republic were launched after the long delays of over a decade following collapse of the communist regimes in Central and Eastern Europe and the formation of the Czech Republic and the Russian Federation as sovereign states. This fact distinguishes Russia and the Czech Republic from the majority of other CEE countries where civil service reform was initiated much earlier.

To a considerable extent implementation of civil service reforms in the Czech Republic and in the Russian Federation was based on the assumption that the reforms constitute mainly legal and organizational (technical) undertakings. In contrast to the Russian Federation, public administration reform in the Czech Republic was viewed until the early 2000s exclusively as reform of territorial government.

Civil service reform agendas in both countries were largely formulated on the basis of information derived from the study of *modern trends in administrative and civil service reforms and other countries' experiences in this field***.** Comprehensive analysis of the problems and shortcomings inherent to the Czech and Russian state administrations was carried out respectively by the Czech National Training Foundation in 1998 and by the group of President Yeltsin's advisors and experts in the same year. However, the results of these analyses were not always appropriately reclaimed later in further work on determining reform's objectives, directions, approaches and results to be achieved.

The reform priorities in Russia and the Czech Republic were marked by *clear orientation towards efficiency and performance***.** Additionally, the reform agendas comprised measures aimed at reorganising central government, introducing new regulatory and institutional framework, creating professional civil service cadre, promoting client orientation, decentralisation and downsizing. Regretfully, orientation towards higher performance and efficiency was not always backed up in either case by consistent practical efforts to reach this objective.

At the same time the civil service reform programmes were distinguished by a certain measure of *eclecticism and compromise***.** This can be explained by weak coordination of donor assistance, general shortage of home based expertise, and influence of corporate and vested interests.

The analysis of the objectives of civil service reform programmes (and of the wider administrative reforms) in the Russian Federation and the Czech Republic shows that these reforms fall mainly into two categories: *structural* (e.g. reorganisation of central government, decentralisation, privatization, size reduction) and *procedural/ technical* (e.g. simplification of procedures, more flexible recruitment, performance related pay). Elements of *budgetary reforms* were also present to a certain extent in the civil service reform programmes but this issue was addressed in a more focused manner within broader administrative reforms.

The reform element of making the bureaucracy more responsive to the public and civil societies, and enhancing transparency and accountability was not sufficiently accentuated and pursued in both Russia and the Czech Republic. Fighting corruption has been also at the periphery of attention of reform managers in practical terms.

It should be remarked that in Russia the civil service reform was implemented at both federal and regional levels, whereas in the Czech Republic the reform encompassed only central state administration.

In terms of pace of reform, we can speak about slow and cautious bordering into a "no movement at all" approach to reform of the central state administration in the Czech Republic (2004 – 2006) and a rather "big bang", "hasty" approach in restructuring the Russian Government in 2004 and relatively more gradual intervention for implementation of the Russian civil service reform programme (2003–2007).

The following features of civil service reforms in Russia and the Czech Republic determined largely their current advance and results:

- the reforms were designed and prepared in a confidential non-transparent manner (no real involvement of citizens, business, non-government organisations and civil servants themselves);
- they have been prepared and implemented in a top-down fashion (orders and instructions) leaving little or no place for "grass root" initiative;
- there has been no real ownership of reform and this complemented by utter shortage of committed stakeholders;
- as a rule criteria of reform effectiveness and achievement of its results have been either vague and general or absent all together leading to a situation when matching them against practice was hardly possible;
- civil society and business community were not invited to monitor the advancement and progress of reform;
- operational management of the reforms appears to have been overall weak.

The main reform instruments turned out to be *project management* in the Czech Republic and *federal targeted programmes* in the Russian Federation. Both instruments have been known for the high emphasis placed on result orientation provided these instruments are appropriately applied.

Typical of the reform approaches in Russia and the Czech Republic has been reliance on *legislation as the main implementation tool*. This rather one sided approach reduced the potential for advance of reforms, particularly in view of a notable discrepancy between civil service law's prescription and manner of enforcement. There has been an obvious "law enforcement" problem in both Russia and the Czech Republic.

Further constraint to reform implementation has been associated with lack of effective management structures guiding and supervising reform implementation and respective law enforcement in the Russian Federation and the Czech Republic. Although formally these structures were created - the General Directorate of Civil Service in the Czech Republic and Commission on Improvement of State Governance in Russia - they could not effectively perform their functions. The former lacked support to its mission and was eventually disbanded in 2006 and the latter turned out to be an ad-hoc structure convened too seldom in order to be able to act as a central point of operational management.

In practice both reforms were not sufficiently focused on rank and file civil servants and were relying mainly on changes driven externally. The reforms largely failed to induce support of reform efforts among the majority of civil servants and politicians or at least to win their loyalty. As a result one could observe either lack of interest at best or passive resistance to reform at worst on the part of the majority of civil servants.

In terms of results of civil service reforms reached so far in Russia and the Czech Republic, it is necessary to underline their major achievement i.e. civil service reforms actually have been launched. This fact entailed a number of positive developments. In particular, the Czech civil service has been finally "legalized" and the Russian civil servants obtained relatively sizable increases to their previously "bare subsistence" salaries.

However, both reforms have been characterized by an overall wide mismatch between intentions and real results so far. Specific improvements in the two civil services and their environment have been so far modest.

With no real ownership and leadership in place and no effective approaches to overcome an existing administrative culture (which remained hostile to the goals and spirit of reform), reform implementation turned out to be in many ways formal and the scope of real tangible changes minimal in practice. It is true that two or three years of reform implementation is not a long enough period to expect cardinal changes to materialize, and even less so to become firmly embedded in the existing structures, habits and practices. However, reform's steady progress and prospects for achieving its main results are largely determined by the existing tendencies in reform implementation, commitment of major stakeholders, support from the politicians and the society, correctness of the selected strategies and methods, sound enforcement and monitoring mechanisms. From these points of view civil service reform capacity in Russia and the Czech Republic continues to be low.

As far as major outputs of civil service reform are concerned they have been identified mainly in the following areas:

- adoption of new legislation and other normative acts (RF & CZ);
- implementation of pilots (experiments) in selected areas, e.g. performance management, result oriented budgeting, greater transparency (RF & CZ);
- wider application of IT technologies (RF & CZ);
- training programmes for civil servants implemented on a greater scale (RF & CZ);
- considerable increase in monetary compensation to civil servants (RF);
- establishment of the High Administrative Court (CZ);
- reduction of administrative barriers for businesses (RF & CZ);
- adoption of standards of quality for services in selected areas (RF & CZ);
- creation of the Institute of Public Administration (CZ);
- specification of the "conflict of interest" in the normative documents (RF&CZ).

For the first time in modern history of Russia experimental introduction of performance related pay took place in 20 government agencies and offices rendering services to citizens and businesses in 2006 (Sharov, 2007). The experiment envisaged considerable bonuses to the staff where the customers were satisfied with the quality of services. Another "innovative" Presidential decree signed in late 2006 allowed the financial benefit from reduction of staff in a public office to be distributed among the best remaining civil servants. (ibid)

In the Czech Republic experiments to enhance quality management – Common Assessment Framework model (CAF) – were carried out in the Office of the Government and the Ministry of Finance in 2006. Information on their impact is not available.

In both Russia and the Czech Republic the reforms have failed so far to introduce changes in the internal incentive mechanism of the civil service and its orientation, as well as to establish effective external control over the service.

The reforms have been also marked by the emergence of certain *side effects and unforeseen results*. In Russia the unforeseen results include (in spite of the reform's declared goals) considerable increase in budget spending on state administration and substantial growth of the number of civil servants at both federal and regional levels in the years 2002–2007. This increase is estimated to be around 20–25% of the entire number of civil servants.

In the Czech Republic unforeseen results were manifested above all in the regular postponement of "validity" of the Civil service Act (except for few provisions) since its adoption in 2002, in the actual suspension of the reform by late 2006 and in the eventual abolition of Civil service Act in August 2007.

9.3 Major Findings of the Research

On the basis of our analysis it may be concluded that the overall need for civil service reform in Russia and the Czech Republic was brought to prominence by the profound transformation of the political and economic systems of these countries in the early 1990s. The decade that followed dissolution of the communist regimes in Central and Eastern Europe witnessed growing contradiction between drastic changes taking place in the political and economic systems of Russia and the Czech Republic and the outdated methods used for managing these countries administratively.

Attempts undertaken in the 1990s to "modernize" the Russian and Czech state administrations failed to eradicate the practices of replacing personal networks of political influence and economic exchange with impartiality, strict rules of professional behavior, and compliance with laws. As in earlier times, a high level of politicization of civil services has been exerting a serious impeding influence on shaping and generating preferred conduct, attitudes, values and objectives of politicians and civil servants in both countries.

The largely uncompetitive political environment that existed in Russia and the Czech Republic throughout most of the 1990s reduced to a minimum opportunities for "correcting" the policies of "little or no interest at all" in launching reform of civil services. In essence the Czech Prime Minister V. Klaus and the Russian President B. Yeltsin (who both were reluctant to start serious reform of state administration) were under no effective pressure (that should usually come from the political opposition, parliament or civil society) to modernize and reform the existing system of state administration.

Ten years after collapse of the communist system the introduction of essential changes into the Russian and Czech state administrations still remained an unfulfilled promise of the political leadership. In these countries many outdated norms and informal rules were still in place, external accountability mechanisms were virtually lacking, corruption and low service capacity remained a serious problem, and transparency was largely discretional. A culture of patronage based relations within which loyalty to the "patron" was the most important prerequisite for a successful civil service career still prevailed. This period also witnessed lack of unity of political and social forces that could introduce and implement genuine civil service reform. Their capacity to influence political decision-making process was also weak.

By the late 1990s pressure for initiation of reform coming from various sources was heightened further by the negative tendencies in the economic and social development of both countries, worsening of practically all parameters of performance of the government and state agencies and, especially in case of Russia, lowering of living standards of the majority of the population. These tendencies had to be taken into account by politicians who wished to count on any respect and support of the electorate.

The civil service reforms in the Russian Federation and the Czech Republic were eventually launched in the early 2000s. Initially, an overall comprehensive approach to reform of state institutions was adopted in both countries, which envisaged combined implementation of a number of reforms - civil service, administrative, regulatory, fiscal management (budgetary), judiciary and others.

Our findings allow conclude that the reform measures undertaken in the two countries, though generally appropriate, have not been sufficiently focused and consistent in addressing the problems of civil service development. Further, the reforms in both countries have been running counter to the prevailing bureaucratic and administrative culture and practices and therefore met with "silent" but overall notable resistance. Finally, support to the reform from the political and administrative elite has been so far largely superficial and conditional. The further the reform implementation proceeded, the less convincing this support appeared to be.

Analysis of the two cases confirmed the significant role played by national leadership in deciding on the issues of initiating and implementing civil service reform. As we could see from the examples of the two countries, there was hardly any advancement in the preparations for reform of civil service during the period when the political leadership was exercised by politicians (B.Yeltsin and V.Klaus), who did not consider such reform a priority.

However, reference to mere absence of political will can hardly help to understand the real causes of lack of or insufficient progress in reforming the Russian and Czech civil services in the 1990s and later. As we emphasized above the institutional framework defines and limits choices available for policy-makers. There should be sound reasons explaining this lack of political will. In search of these reasons it is critical to study the main characteristics of the civil services in question. It is no less important to explore the principal features of the political and economic setting as well as peculiarities of existing mentality, traditions, culture and historical development.

Of primary importance here are low extent of political competition among various parties and coalitions and the high level of politicisation of the Czech and Russian state administrations noted by many experts and confirmed by the results of surveys amongst civil servants in both countries. For obvious reasons the Czech and Russian state administrations were predominantly politicized during communist rule. But little has changed in this respect after "perestroika" and the "velvet revolution".

Identification of key characteristics of the civil service systems in question through the application of framework research protocol for comparative studies allowed us to conclude that the basic features of merit based civil service systems (appointment and promotion based on open competition, impartiality and performance based remuneration, developed system of job descriptions etc) were largely absent in both Russia and the Czech Republic. In particular we pointed to wide discretionary power of superiors in deciding on civil servants' promotion as well as in determining the size of take home pay of their subordinates.

The persistent elements of the patrimonial bureaucracy present in both civil service systems constituted a serious obstacle to successful implementation of civil service reforms. The situation was further compounded by the weakness of civil service external accountability mechanisms in both countries.

The analysis carried out has enabled us to find answers to the research questions formulated at the very start of the research. The main research question was defined as follows:

What have been the experiences with the preparation, launching and implementation of national civil service reform in the Russian Federation and the Czech Republic following the fall of communist regimes and which factors can explain the relative success or failure of the reform proposals.

After lengthy and difficult periods of delay and uncertainty, civil service reforms in the Russian Federation and the Czech Republic were eventually launched in 2003-2004. Out of a whole number of reasons explaining these delays and difficulties two major interrelated factors have served as key constraints in launching civil service reform: a) high level of politicization of both civil services complemented by widespread clientelistic and patronage networks; and b) seriously weakened political competition between main political parties/coalitions both in the Czech Republic and Russia.

Notable factors, differing in each country, played a decisive role in triggering civil service reforms. In the Czech case it was permanent pressure applied by the European Commission in view of the planned accession of the Czech Republic to the European Union. In the Russian case it was the election of a new President who committed himself to reform the state bureaucracy.

There has been a general tendency in both Russia and the Czech Republic to maintain confidentiality around reform preparation efforts and to view civil service reform as a primarily legal and technical matter. Within the framework of the launched reforms introduction of the principles of meritocratic, efficient, performance oriented, accountable and corruption immune civil service has not been sufficiently consistent and result oriented both in Russia and the Czech Republic. Hidden resistance to reform has been noted at practically all levels of state administration.

In view of rather limited time that has passed since the start of reforms one can speak only of intermediary results achieved by the reformers. These appear to be rather modest.

Within the framework of the preparation and implementation of reform of the Czech central state administration the results achieved so far are manifested mainly in: the adoption of Civil service Act; creation of the Institute of Public Administration; elaboration of methodological approaches for introduction of the planned innovations in civil service and HR management; pilot application of CAF model in central government; expanded programmes of civil servants' post-entry training; and wider IT application in the Czech ministries.

By 2007 civil service reform in the Czech Republic had been virtually put on hold. This fact was largely pre-determined by a combination of factors including: overall initial low capacity for civil service reform; inertia and resistance of political and bureaucratic elites fearing that upcoming changes would undermine their privileged position and existing clientelistic networks (right for "spoils"); administrative traditions and culture rooted in the communist and pre-communist past; failure to win broad support for reform from civil servants, businesses and civil society; and virtual cessation of the pressure coming from the European Commission following accession of the Czech Republic to the European Union in 2004.

By 2007 the civil service reform in Russia appeared to be advancing further albeit at a slower rate with a number of principal outcomes and outputs still outstanding. Major up to date achievements were confined mainly to the adoption of new civil service legislation; restructuring of the government (as part of administrative reform); development and adoption of modern job descriptions; implementation of pilots in such areas as introduction of performance based pay and standards of services; elaboration of methodological tools for introducing other planned innovations; and increased remuneration of civil servants.

In spite of certain setbacks in reform implementation the civil service reform remains an important issue of the government agenda in Russia. The driving force of the reform continues to be the Russian President who in 2005 extended implementation of the civil service reform programme for another two years. Draft programme of civil service reform and development for the period 2008-2013 has been submitted for consideration of the Russian Government.

The prospects for success of civil service reform in Russia depend to a considerable extent on the continuity of the reform policies after completion of the President Putin's term in office in 2008. Should the reform be suspended or fail to reach its major objectives, the current trends in the country's political and economic development may lead to eventual 'privatisation of the Russian state by its bureaucracy'.

Our research led to the identification of answers to research sub questions, which are provided below:

1) What were the main characteristics and prevailing features of each of the two civil service systems (Russian and Czech) before the reform was designed and launched?

During most of the 1990s the civil service systems in the Russian Federation and the Czech Republic were characterized by: the persistence of the elements of patrimonial bureaucracy to the detriment of wide application of meritocratic principles; domination of informal clientelistic, patronage networks; weak external accountability; low efficiency; discretional transparency; and increasing level of corruption. The combination of these features complemented by excessive politicization of civil services, weakness of appropriate regulatory and institutional framework, shortage of relevant expertise and tight fiscal constraints determined overall low reform capacity of the Russian and Czech civil services.

2) What are the dynamics of change and contents of reform of the Russian and Czech civil services?

It had taken over a decade since the demise of the communist system in the two countries before some real steps to implement effective civil service reform were made. Reform agendas in both countries were largely formulated on the basis of information derived from the study of modern trends in administrative and civil service reforms and other countries' administrative experiences (mainly EU members).

Civil service reforms in Russia and the Czech Republic have been distinguished by their focus on enhancing efficiency and performance which have been declared the principal objectives of the reforms. Overall the reform agendas in both cases prioritized the *management* tradition (emphasis above all on efficiency and economic rationality). Considerably less attention has been paid to the *governance* tradition (treatment of civil service as primarily a mechanism for effective governability and political legitimacy). The analysis of the reform programmes and new legislation also allows us to conclude that general orientation of civil service reform in Russia has been towards creating a career (tenure) system, whereas the reform in the Czech Republic has been geared to build up a mixed system (position/career).

The contents and main directions of the reforms comprised a certain measure of eclecticism and compromise, explained above all by sheer shortage of national expertise, weak coordination of donor assistance, and strong impact of vested and corporate interests.

One of the major paradoxes of civil service reforms in Russia and the Czech Republic was that although practical introduction and application of performance and results oriented management were declared as one of the principal objectives of administrative transformation, in managing the reforms themselves these objectives were *de facto* largely ignored. Out of a variety of available implementation tools exclusive preference was given to legislation. Other tools such as training and best practice diffusion were applied but in a rather fragmented and not always consistent manner. Fiscal incentive instruments, networking and coalition building have been practically disregarded.

The reform objectives of making the bureaucracy more responsive and accountable to the public and civil society, distinct in many EU countries, have not been sufficiently accentuated and appropriately addressed yet in the Czech and Russian cases. Little attention, if any, has been paid to the issues of capacity building and reducing opportunities for corruptive behaviour.

Intensity of reform policies and efforts varied between the two countries at different stages of their development. At the start up stage we can speak about slow and cautious, bordering to "no movement at all" approach to reform of the central state administration in the Czech Republic (2004 – 2005). This is in comparison to a "big bang", "shock like" approach in restructuring the Russian Government in 2004 with a relatively more gradual intervention as far as subsequent changes in the Russian civil service itself are concerned (2004 – 2007).

3) What are the essential factors that influenced the pace and current results of reform?

The analysis confirmed significant impact of the existing institutional setting and legacies on the process of civil service reform. As we have shown, throughout the 1990s the prevailing institutional environment in the Russian Federation and the Czech Republic was characterized by weakened political competition, entrenched clientelism, excessive politicization, immature civil society, and shortage of funding and expertise. All these factors considerably reduced capacity for reform of the Russian and the Czech civil services.

Under such conditions the driving momentum for launching reform could come either from outside pressure or from a charismatic leader. Eventually this is what happened in the Czech Republic where major factor to launch and sustain reform was permanent pressure applied by the European Commission; and in Russia where the main driving force is associated with the President elected in 2000 and who committed himself to reform the state bureaucracy.

Apart from low reform capacity described above the pace and results of the reforms have been affected by other factors. In particular, the reforms suffered from actual absence of committed reform stakeholders and hence lack of real reform ownership. Reforms were predominantly viewed as something imposed externally, either by top authority (the President in the Russian case) or by a foreign community (European Commission in the Czech case). Lack of ownership was also manifested in the shortage of leaders who could move beyond traditional perceptions of management and display resoluteness and consistency in driving innovative approaches.

Reform implementation was also influenced by unpredictability and uncertainty of political and economic situation in both Russia and the Czech Republic (e.g. frequent change of governments, political and economic crisis, unstable coalitions, etc).

Most importantly, in both Russia and the Czech Republic the reforms failed so far to introduce changes in the internal incentive mechanism of the civil service and its orientation, as well as to establish effective external control over civil service. The reforms neither succeeded in attracting broad support from civil servants, citizens and prominent politicians, nor made them at least loyal to the reform policies. Civil society, public organizations and businesses have not been appropriately involved in the preparation and implementation of the reform and hence neither provided required backing, nor acted as effective monitoring and control mechanism of the reform advancement. These factors reduced considerably the possiblities of effective and timely implementation of the reforms' programmes.

The weakest point of reform management and strategic approach in both cases was excessive and even exclusive reliance on legislation as the main instrument of reform implementation and absence of permanent and effective management structures enjoying broad support to their mission. This situation was further compounded by the problem of weak law enforcement in both countries.

4) How can the differences and similarities in the development and reform of the two civil services be explained in order to define possible causes of similarities and variations?

Apart from weak reform capacity which determined the existence of many commonalities in the reforms of the Russian and the Czech civil services, similarities in reform preparation and implementation are in particular attributed to the largely common roots of administrative culture and mentality. The latter can be traced to the heritage of monarchist (Austro-Hungarian and Russian Empires) and recent communist regimes. The political and administrative systems in the former communist countries constituted the extreme case of the development of uncontrolled and unaccountable bureaucracy. The political regime in Czechoslovakia before the "velvet revolution" was one of the most orthodox and ideologically rigid in Eastern Europe. In this respect it was very similar to the political regime in the Soviet Union. This similarity has been one of the important factors explaining many common traits of civil service development and reform in both countries.

In the early 1990s both states underwent the so called "triple transformation" - a unique form of de-institutionalization in the politics, economy and nationhood which distinguished them from most other post-communist countries. In addition, the reformers in both Russia and the Czech Republic have been using largely the same sources of inspiration (in particular, experience of advanced EU member states) in the elaboration of the reform programmes.

In spite of a considerable number of similarities the reforms in the Russian Federation and the Czech Republic were distinguished by a number of significant differences. The variance is manifested above all in the major driving forces of reform (European Commission in the Czech case and new President in the Russian case) and in the current status of reforms. Contrary to the situation in the Czech Republic where the declared process of accession to EU entailed a considerable pressure on the Czech authorities from Brussels to implement genuine civil service reform there was no such external pressure experienced by the Russian authorities.

The causes of major differences between the two countries in the civil service reform processes and current outcomes (above all suspension by 2007 of the reform in the Czech Republic and continuation of the reform in the Russian Federation) are largely rooted in the distinctions in the existing institutional configurations of the two states. As we have shown in our study, this difference in particular stems from: relatively greater initial civil service reform capacity in Russia (e.g. legal existence of civil service in Russia since 1993 and in the Czech Republic only since 2002); major driving forces for reform (European Commission which virtually ceased its pressure after the Czech Republic joined EU; and the RF President who committed himself to reform the Russian bureaucracy); peculiarities of the constitutional framework and political regime (in the Czech Republic - rotating political coalitions concerned to preserve the right to "spoils"; in the Russian Federation - the President with relatively more powers in enforcing his policies and much less dependent on the narrow interests of political parties); and finally lack of commitment of the Czech political elite to civil service reform.

The early 2000s saw the emergence of another distinction affecting civil service reforms in Russia and the Czech Republic – continuing budgetary constraints in the Czech Republic and notable relaxation of fiscal pressures on the reform in Russia resulting mainly from the considerable increase of oil and gas prices on the world markets. Growing budgetary revenues in Russia facilitated continuation of civil service reform (the reform in the Czech Republic had no such benefits), though this factor appears to have contributed to the emergence of certain side effects, in particular, marked growth of the number of civil and municipal servants and large increase in the allocation of state budget funds to support the state bureaucratic establishment.

Our analysis has demonstrated that economic and social development of post-communist countries is contingent upon the process of successful reformation of their state administrative systems.

The research findings have illuminated the problems that continue to impede or delay civil service reforms in the Russian Federation and the Czech Republic. Understanding of these problems is important for the identification of the appropriate remedial actions and their consistent application in practice.

The research conclusions suggest important lessons for the reformers in those post-communist countries which have embarked on the road of transformation of their civil services and would like to avoid mistakes and difficulties experienced by their Russian and Czech colleagues.

The existence at present of more favourable conditions for continuation of civil service reform in the Russian Federation does not necessarily mean that in the medium and long-term perspective the demands for administrative change in Russia will continue to be sufficiently strong and that the reform in the end is likely to be more successful than in the Czech Republic. The experience shows that the reforms, which depend mostly on personalities, are not sufficiently sustained and may be put on hold when a new leader comes to power. In this respect the prospects for successful implementation of civil service reform in the Czech Republic, which is obliged to conform to EU standards, look in the long term more favorable than in Russia.

9.4 Theoretical Conclusions and Implications

Apart from specific findings linked to the particular research question and sub-questions our comparative study allows us to draw certain conclusions as far as theoretical and methodological approaches and assumptions in comparative research of administrative systems are concerned.

To illuminate these conclusions it is worth returning shortly to the start of the study and recalling its basic methodological cornerstone – the institutional analysis. As stressed more than once in chapters I and II and also later, the institutional analysis implies that development of civil service system should be studied above all from the point of view of existing traditions, rules, culture and authority relationships which persist in time and exert significant influence on the evolution of the entire civil service system, civil servants' conduct and attitudes, and prospects of civil service reform.

Having finalized our study we may confidently argue that it has reaffirmed the basic postulate i.e. in order to understand processes of civil service reform, distinctions and similarities between reforms in different countries "...we need to search for explanatory factors, not only directly given from within civil service systems as such, but also – and probably mainly – derived from the political and societal environment of the civil service system in question" (Bekke, 1999: 11).

In particular, the study has demonstrated that problems and constraints in the development of civil service systems are largely rooted in specific political arrangements, social norms, traditions and patterns of behaviour rather than administrative structures. The studied cases show for example, that cardinal re-structuring of the Russian Government in 2004 has not led to any noticeable increase in efficiency of state administration, shift in attitudes of civil servants or their greater accountability. Furthermore, subsequent developments disclosed certain retreat from at least some of the declared principles of the reform of the Russian Government.

This and many other examples taken from our study point to an enduring impact of traditions and path dependence on the processes and contents of civil service reforms. The analysis reaffirmed, in particular, remarkable capability of the state bureaucracies in these countries to withstand large scale transformation in the political, economic and social spheres. Our study has shown that informal rules, traditions and behavioral patterns rooted in the communist and pre-communist past have been constantly reproducing themselves in the new environment. As an example we can refer to the deeply entrenched habit of the Russian and Czech political elites to treat the state institutions and civil service as the instrument of their personal power.

The influence of path dependence largely explains the persistence of irrational and outdated systems (in our case administrative systems in post-communist countries) that have been rendered obsolete by new social, economic and political circumstances in societies undergoing rapid and profound modernization.

However, that does not mean that we should view the path dependence concept as an instrument of re-inventing a notion of causality in administrative science. Our analysis can conclude that a particular path of civil service reform and success or failure of reform strategies have been the result of impact of a combination of institutional and leadership factors. We also acknowledge that contingent events may play a distinct role in the development and reform of administrative systems.

The study demonstrated that differences and similarities in civil service reform processes and results originate largely from variance and commonalities in the key elements of institutional architecture in the countries concerned. In this connection the study also affirmed the importance and usefulness of the concept 'capacity of civil service reform' and its three major parameters (widely used in our study). The concept and parameters provide general guidelines for assessing the reform potential of a given country and explaining successes and failures of reform processes.

The research revealed that tradition appears to have played so far mostly a restrictive role in the reform of the Russian and Czech state administrations serving as an impediment to change (e.g. legalistic approach to reform, wide-spread patronage networks, etc) and not leading to change. However, at least one case demonstrated a somewhat positive (though temporary) role of tradition - tradition of a strong centralised power in Russia seems to have facilitated launching and powering civil service reform by a new President. This notwithstanding, the same tradition may have also negative implications as it eventually constrains local initiative and discourages a "bottom up" approach.

The study showed that there are more similarities than differences in the reforms of the Russian and Czech civil services. In other words the conclusion made over 15 years ago that "...all the same, public administration across these (post-communist-A.K.) countries is more notable for similarities than differences both in its shortcomings and the stages of reform" (Rice, 1992: 117) retains its validity.

The study allowed the tracing of an interesting dependency – the higher world oil prices and hence revenues of the state budget are, the more the Russian state administration expands and consumes resources. There are sufficient grounds to assert that this phenomenon may also emerge in a similar situation in other countries where there exist insufficient external accountability of bureaucracy, weak enforcement of meritocratic principles, high level of politicization of the state administration and wide spread corruption.

The processes of reform preparation and implementation in the selected post-communist countries witnessed at times dramatic episodes. It is enough to recall that the Civil service Act was pushed through the Czech Parliament by a majority of only one vote and following numerous postponements of full enforcement was abolished all together five years later. In Russia the full text of the concept of civil service reform was kept confidential for over a year after approval by the Russian President. In one interview President Putin admitted that the main lesson he had learnt was "the extreme difficulty of fighting bureaucracy" (The Observer, 8 June 2003). These and many other examples from this study provide enough evidence of the existence of the forces of strong opposition to civil service reform in each of the countries concerned.

Finally, the findings of our study demonstrate that the major problems of civil services in post-communist countries are not so much those of low efficiency and effectiveness, weak legal framework and poor quality of rendered services (though they all are important). More significant are the problems directly linked to the relations and interface of the state bureaucracies with the political, societal and cultural environment i.e. genuine commitment to democratic values and service to citizens, genuine impartiality and accountability, high ethical and professional standards. Effective remedies to treat these "diseases" are well known: above all developed and well functioning democratic institutions, mature civil society and strict abidance by the rule of law. However, as our analysis of the two post-communist states demonstrates it is rather easy to prescribe a "medicine", whereas it is much more difficult and problematic to apply it in an appropriate and sustained fashion adjusted to a specific situation.

Bibliography

Aarrevaara, T., (1999) 'Restructuring Civil Service in Russian Public Administration', Helsinki: Finnish Ministry of Finance;

Aberbach J. and Rockman B., (1992) 'Does Governance Matter ? And If So, How?' *Governance* , Vol. 5 , No 2, pp. 135-153;

Adam, J., (1999) 'Social Costs of Transformation to a Market Economy in Post-Socialist Countries', New York, NY: St. Martin's Press, 1999;

Afanasyev, M., (1998) 'Clientelism and the Russian Statehood' (in Russian), Moscow: Moscow Scientific Foundation;

Akmiska, M., (2000) 'The Czech Multipartism in the late 1990s,' *Central Europe Review*, N2, 2000;

Almond, G., (1966) "Political Theory and Political Science", *American Political Science Review*, 60, pp 869-79;

Arbatov, A., (2007) 'Bureaucracy On the Rise', *Russia in Global Affairs,* N 2, Moscow;

Atamanchuk, G. (2002) 'The Essence of Civil Service' (in Russian), Moscow: The Russian Academy of Civil Service;

Baker, R., (ed) (1994) 'Comparative Public Management', Westport: Praeger Publishers;

Behlavy, M. (2002) "Management of Civil Service Reform in Central Europe" in Peteri G. (ed), *Mastering Decentralization and Public Administration Reforms in Central and Eastern Europe* OSI.LGI Budapest, pp. 55-71;

Bekke H., Perry J. and Toonen Th. (eds), (1996) 'Civil Service Systems in a Comparative Perspective', Bloomington: Indiana University Press;

Bekke H., (1999) 'Studying the Development and Transformation of Civil Service Systems: Processes of De-institutionalization', in: Jay D. Whyte & James L. Perry (ed.), *Research in Public Administration*, vol. 5, Jai Press Inc. Stamford, pp. 1-18.;

Bekke H. and van der Meer F., (eds), (2000) 'Civil Service Systems in Western Europe', Cheltenham: Edward Elgar;

Belohradsky V., (2000) 'Realism in Politics Worries People', *Central Europe Review* Vol. 2, No 20;

Bevir M., Rhodes R. A. W. and Weller P., (2003) "Traditions of Governance: Interpreting the Changing Role of the Public Sector in Comparative and Historical Perspective". Public Administration. 81, pp. 1-17;

Boin, R. and Christensen, T., (2004) 'Reconsidering Leadership and Institutions in the Public Sector. A Question of Design?', Leiden University;

Boone, P. and Horder J., (1998) "Inflation: Causes, Consequences, and Cures", in Peter Boone, Stanislaw Gormulka an Richard Layard (eds*) Emerging from Communism: Lessons from Russia, China, and Eastern Europe,* Cambridge: MIT Press;

Campos, N., (1999) "Context is Everything: Measuring Institutional Change in Transition Economics", Prague: CERGE-EI;

Cebisova, T., (1996) 'Civil Service Reform in the Czech Republic' in '*The State, Problems and Outlines of Public Administration in the Czech Republic*' OLPA, Prague;

Chandran P., (2001) 'The Czech Republic', GOPAC;

Chandler, J.A. (ed), (2000) 'Comparative Public Administration', London and New York: Routlegde;
Chaudhry S., Reid G. and Malik W., (eds) (1994) 'Civil Service Reform in Latin America and Caribbean', Proceedings of a Conference, Washington: World Bank;
Collection of Laws of the Russian Empire, (1830), Vol. 1/V, N 3393, St Petersburg;
Concept of Modernization of Central State Administration Considering the Status of Civil Servants and Structure of Administration Authorities, (2001) Resolution of the Czech Government No. 619 of 20 June 2001, www.vlada.cz ;
Concept and Main Directions of Reform and Modernization of Central State Administration, (2004) Resolution of the Czech Government No. 237 of 17 March 2004, www.vlada.cz ;
Concept of Training Civil Servants (1999) Resolution of the Czech Government No. 601 of June 24, www.vlada.cz ;
Coombes, D., (1997) 'Alternative Paths for the States of Central and Eastern Europe', in Coombes, D, and T. Verheijen, (eds), *Public Management Reform, Comparative Experiences from East and West*, Plovdiv: Litera, pp. 401-407;
Coombes, D., (2002) 'Public Administration Reform as Deconstruction of the State', Paper, presented to NISPAcee conference in Riga, Latvia;
Cummings, S. and Nørgaard, O., (2003) 'The Administration-Society Interface. A Comparative Perspective on Post-communist Experiences'. Paper prepared for the 11th NISPACee Annual Conference 'Enhancing the capacity to govern. Challenges facing the CEE Countries'. Bucharest, Romania, April 10-12 2003;
Čulík, J., (1999) 'A Fistful of Impressions'. *Central Europe Review.* Vol.1 No.23, 29 November 1999;
Čulík, J., (2000) 'Anti-Communism and Emotional Substitutes: the Forms of Legitimacy in the Democratic Construction,' Prague: CEFRES;
Dahl, R., (1971) 'Polyarchy: Participation and Opposition', New Haven: Yale University Press;
Debates in the Czech Parliament, 12 April 2002, Verbatim report, http://www.psp.cz/cgi-bin/ascii/eknih/1998ps/stenprot/047schuz/s047009.htm ;
Delyagin, M., (2002) "A Year of Ruined Hopes: Source of Optimism", Moscow, Vlast, N3, 2002;
Delyagin, M., (2005) Speech in the city of Pskov on 14 July 2005, www.delyagn.ru;
Dmitriev, M., (2002) Speech at the meeting in the Office of Carnegie Endowment for International Peace on 1 May 2002. www.csr.ru;
Dmitriev, M., (2004) Interview to Profil magazine N 44, 6 December, 2004;
Dimitrova, A., (2002) 'Enlargement, Institution-Building and the EU s Administrative Capacity Requirement' , *West European Politics*, 25 (4), pp. 171-190;
Drengsgaard, L., (2003), 'A Game Theoretical Analysis of the Hungarian Civil Service System', Working Paper, NISPAcee-Annual Conference, Bucharest;
Dvorakova, Z., (2005) "HR Practices in the Czech Public Administration", paper prepared for the European Group for Public Administration conference in Bern 31 August – 3 September 2005;
Evans, P. and Rauch, J., (1999) 'A Cross National Analysis of the Effects of "Weberian" State Structures on Economic Growth', *American Sociological Review*, Oct 1999; Vol. 64, no 5.

Fernandez J. G., (2001) 'Czech Republic: Constitutional Framework', Report to Phare project N CZ 9808.01;
Farazmand, A., (ed.) (1994) 'Handbook of Bureaucracy', New York: Marcel Dekker, Inc.;
Federal State Statistics Service, (2005), Statistical bulletin N 8 (118), Moscow;
Federal Programme "Reform of the Civil Service of the Russian Federation (2003-2005)", (2002) Moscow: Infra M Norma;
Fried, R., (1990) "Comparative Public Administration: The Search for Theories". In Lynn N. and Wildavsky A., (eds.), *Public Administration: The State of the Discipline*. Chatham, New Jersey: Chatham House Publishers, pp. 318-44;
Further Advance of Reform and Modernization of Central State Administration in the Years 2005 - 2100, (2005), a Document (N 02958/05–KVÚ of 21 February 2005) submitted to the Czech Government, Praha: www.reforma.vlada.cz;
Gabrelian, V., (1999) "Post-Communist Bureaucracies: Organisational Modes of Transition" *International Journal of Public Administration*, Vol. 22, No.1, pp.39-68;
GfK Praha (2001), Press release, 7 August 2001, www.gfk.cz/corruption ;
Gimpelson, V. and Magun V. (2004) 'Perspectives and Limitations in the Career of Young Civil Servants', (in Russian) Moscow: High School of Economics;
Goetz, K. (ed.), (2001a), 'Executive Governance in Central and Eastern Europe', special issue of *Journal of European Public Policy*, 8 (6), 2001;
Goetz, K., (2001b) 'Making Sense of Post-Communist Central Administration: Modernization, Europeanization or Latinization?' , *Journal of European Public Policy* 8 (6), S. 1032-1051;
Goetz, K., (2003) 'Executives in Comparative Context' , in J. Hayward and A. Menon (eds.) *Governing Europe. Memorial Volume for Vincent Wright,* Oxford: Oxford University Press;
Goodin, R., (1996) 'The Theory of Institutional Design', Cambridge: Cambridge University Press;
Gray, P., (2003) 'Hard Cases and Improving Governance: Putin and Civil Service Reform in the Russian Federation', Paper presented at the 11th NISPAcee Annual conference: Enhancing the capacities to govern: challenges facing the CEE countries. Bucharest 10th-12th April, 2003;
Grishkovets, A., (2002) 'The Problems of Legal Regulation of the Civil Service in the Russian Federation', (in Russian), Moscow: Institute of Upgrading Civil Servants' Skills;
Grzymala-Busse, A., (2003) "State Reform and Informal Practices in East Central Europe", Yale University;
Grzymala-Busse, A., (2004) "Post-Communist Competition and State Development", Working paper, University of Michigan;
Grzymala-Busse, A., (2001) 'Informal Institutions and the Post-Communist State', Department of Political Science, Yale University;
Hague R., Harrop M., Breslin S. (1992) 'Comparative Government and Politics'. New York: St. Martin's Press;
Halligan, J., (ed), (1998) 'Public Sector Reform', Canberra: Center for Research in Public Sector Management', University of Canberra;
Halligan, J., (ed), (2004) 'Civil Service Systems in Anglo-American Countries', Cheltenham: Edward Elgar;

Hallstrom, L. K. (2005) "Democracy From Above: ECE Environmental Policy Elites and European Integration", Paper presented at the First European Conference of the International Society for Third-Sector Research (ISTR), Paris, France. April 27-29, 2005;

Hanley S., (1999) 'Bureaucracy Is as Czech as Beer', Central Europe Review N 26, http://www.ce-review.org/authorarchives/hanley_archive/hanley26old.html ;

Havel V., (1997) State of the Nation Address to a joint session of the Czech Parliament in Prague, 9 December, 1997, http://www.psp.cz ;

Hesse J.J., (ed.) (1993) 'Administrative Transformation in Central and Eastern Europe: Towards Public Sector Reform in Post Communist societies', Oxford: Blackwell Publishers;

Hesse J.J., (1993) 'From Transformation to Modernisation: Administrative Change in Central and Eastern Europe', in *Public Administration* N 71 (1-2) March 1993, pp. 219-257;

Hesse J.J., (1995) "Comparative Public Administration: The State of the Art" in Hesse J. and Toonen T. (eds) 'The European Yearbook of Comparative Government and Public Administration', Baden Baden/Boulder: Westview Press;

Hesse J. J., (1998) 'Rebuilding the State: Administrative Reform in Central and Eastern Europe', in *Preparing Public Administrations for the European Administrative Space*, SIGMA Papers N 23: SIGMA-OECD (pp 168-177), Paris;

Hesse J.J., Hood C. and Peters B.G. (2003) 'Paradoxes in Public Sector Reform: An International Comparison', Berlin: Duncker and Humblot;

Heady F., (1998) 'Comparative and International Public Administration: Building Intellectual Bridges', *Public Administration Review* January/February 1998, Vol.58, No.1, p.32-39;

Heady, F. (1996) 'Configurations of Civil Service Systems', in Bekke H., Perry J. and Toonen Th. (eds), *Civil Service Systems in a Comparative Perspective*, Bloomington: Indiana University Press, pp. 207-226;

Heady, F., (2001) 'Public Administration. A Comparative Perspective', New York: Marcel Dekker;

Hough, J., (1973) 'The Bureaucratic Model and the Nature of the Soviet System', *Journal of Comparative Administration,* 5 No.2;

INDEM, (2002) 'Diagnosis of Corruption in Russia: Sociological Analysis', (in Russian), Moscow: INDEM Foundation;

Innes, A., (2002) 'The Changing Power of the State in Eastern Europe', Paper presented at the ECPR Workshop 'Enlargement and European Governance', Turin, 22-27, March 2002;

Jalovecká, M., (2005) Political and Administrative Roles of the Office of the Government of the Czech Republic in the Policy-making Process, Praha;

Jenkins, R., (1997) 'The Professionalization of the Public Sector and the Implications for Democratization in Central and Eastern Europe'. Policy Paper, International Research & Exchanges Board. Washington, D.C.;

Jreisat J., (2002) 'Comparative Public Administration and Policy', Westview Press;

Jreisat J., (2005) 'Comparative Public Administration is Back In, Prudently' in *Public Administration Review*, Vol. 65, N 2, pp. 231 – 242;

Johannsen, L., (2004) "Power and Coordination: Decision Making and Influence in Post-communist Governments", Paper prepared for ECPR Joint Workshop *The Process of Decision Making in Cabinets in Central – Eastern and Southern Europe.* Uppsala, Sweden, April 13-18 2004;

Johannsen, L., and Nørgaard O., (2004) 'Governance in Central and Eastern Europe: A cross-sectional perspective'. *How the EU accession broke institutional path dependencies in post-communist regimes.* By Department of Political Science University of Aarhus Denmark;

Jones, L.R. and Kettle D.F., (2004) 'Assessing Public Management Reform Strategy in an International Context', in Jones L., Schedler K., and Mussari R.,(ed) *Strategies for Public Management Reform,* New York: Elsevier Publishers;

Kamenka, E., (1989), 'Bureaucracy'. UK: Blackwell;

Kampars N. A. (2004) "The Troubled Path of the Transformation Process in the Czech Republic", Central Europe Review, N 4;

King V., (2003) 'What Kind of Civil Service? Trends in Public Administration Reform in Eastern Europe', in *NEBI Yearbook 2003*, Springer., pp. 55-66;

Konopasek Z., Kusa Z., Stockelova T., Vajdova T., Zamkalova L., (2002) 'Public Accountability: Czech National Profile', Research report, European Research Project, CT2001-00076. Praha, manuscript;

Knight, J., (1992) 'Institutions and Social Conflict', New York: Cambridge University Press;

Kotchegura A., (1999) 'A Decade of Transition Over: What Is On the Administrative Reform Agenda?' in Verheijen T. (ed) *The Civil Service Systems in Central and Eastern Europe*, London, Elgar Publishers;

Kotchegura A., (1999) 'The Russian Civil Service. Legitimacy and Performance' in Verheijen T. (ed) *The Civil Service Systems in Central and Eastern Europe*, London: Elgar Publishers;

Kotchegura A., (1999) "International Assistance to the Modernisation of the Russian Public Administration - Towards Greater Co-ordination and Sustainability", in *NISPAAcee News,* Vol. VI, No 3, Bratislava;

Kotchegura A., (2002) "Administrative Reform: Problems and Prospects of Transformation of the Russian State Administration", Policy paper, Russian European Centre for Economic Policy, Moscow: RECEP;

Kotchegura A., (2004) "Reducing Corruption in Post-Communist Countries", in *Strategies for Public Management Reform*, New York: Elsevier Publishers;

Kotchegura A., (2006) 'Post – Entry Training of the Russian Civil Servants: Issues and Solutions' in *Civil Service Reform*, Moscow, Lenand Publishers;

Kotchegura A., (2008) 'Professional Development Programmes As a Key Instrument of Enhancing Civil Servants' Performance' (in Russian) in *Vestnik RPFU 'Sociology'*, Moscow: RPFU Publishing House;

Kulinchenko A., (2002) 'Contemporary Problems of Modernization in Russia', in *Vlast,* No. 1, Moscow;

Landová M., (2006) 'Trimble Legal Landscape in a "Legislative Hurricane:" Impact of the 2004 European Union Enlargement on Czech Legislative Process', Stanford Law School;

Lane J-E. (ed), (1997) 'Public Sector Reform: Rationale, Trends and Problems'. California: Sage;

Larjavaara, I., (2001) 'The Institutional Foundations of Administrative Development in Russia', HAUS (Finnish Institute of Public Management), www.helsinki.fi/aleksanteri/english/publicaions/ap_2-2001.pdf;
Liegle B, and Muller W. (1999) "Senior Officials in Austria" in E.Page and V.Wright (eds) *Bureaucratic Elites in Western European States*, New York: Oxford University Press;
Lizal L. and Kocenda E., (2002) 'The Paradox of Czech Crusaders: Will They Ever Learn the Corruption Lesson?', CERGE-EI, Prague;
Lynn B. and Wildawsky A., (ed.) (1990) 'Public Administration. The State of the Discipline'. New Jersey: Chetham House Publishers, Inc.;
Marek J., (1999), Speech of the Head of Public Administration Reform Department, Ministry of the Interior of the Czech Republic in Maastricht in October 1999;
McGregor E. and Solano P., (1996) 'Data Requirements and Availability' in Bekke H., Perry J. and Toonen Th. (eds), *Civil Service Systems in a Comparative Perspective*, Bloomington: Indiana University Press, pp. 42-64;
Manning N. and Parison N., (2004) 'International Public Administration Reform: Implications for the Russian Federation', Washington DC: World Bank;
March J.G. and Olsen J.P. (1984) 'The New Institutionalism: Organisational Factors in Political Life', *American Political Science Review 78* (September) pp. 734-749;
March J.G. and Olsen J.P. (1989) 'Rediscovering Institutions: The Organisational Basis of Politics', New York: Free Press;
Maslovsky M., (1997) 'Theory of Bureaucracy of M.Weber and Modern Political Sociology" (in Russian), Nizhny Novgorod: N/Novgorod State University Print house;
Meer Frits M van der and Roborgh Renk L (1996) 'Civil Servants and Representativeness' in Bekke H., Perry J. and Toonen Th. (eds), *Civil Service Systems in a Comparative Perspective*, Bloomington: Indiana University Press, pp. 119-136;
Meer Frits M van der, Raadschelders Jos C.N. and Toonen Theo A.J. (eds.) (2007) 'The Civil Service in the 21st Century: Comparative Perspectives' Houndsmills: Palgrave/McMillan;
Meer Frits M van der, Steen Trui and Wille Anchrit, (2007) 'Western European civil service systems: a comparative analysis' in: Frits M van der Meer, Jos C.N. Raadschelders and Theo A.J. Toonen (Eds.) (2007), 'The Civil Service in the 21st Century: Comparative Perspectives' Houndsmills: Palgrave/McMillan;
Meyer-Sahling J., (2006) 'The Institutionalization of Political Discretion in Post-Communist Civil Service Systems: The Case of Hungary'. *Public Administration*, 84(3), pp. 693-716;
Meyer-Sahling, J. (2004) 'Civil Service Reform in Post-Communist Europe: The Bumpy Road to Depoliticisation'. *West European Politics*, 27(1), pp. 69-101;
Meyer-Sahling J., (2002) 'Personnel Policy Regimes, Political Discretion and Civil Service Reform in Central and Eastern Europe', Paper prepared for Workshop 4, 'Enlargement and European Governance', ECPR Joint Session of Workshops, Turin/Italy, March 22-27, 2002. London School of Economics and Political Science, Department of Government;

Metcalfe Les, (1998) 'Meeting the Challenges of Accession in Preparing Public Administrations for the European Administrative Space', SIGMA Papers N 23: SIGMA-OECD (pp 41-62), Paris;
Mišovič, J. and Tuček M., (2003) "How the Czech Public Views the Political and Economic Elites", Praha: Sociological Institute of the Czech Academy of Sciences;
Moller L. and Norgaard O., (2004) 'Who said politicians cannot be conservative? Comparing reform capacity in the Czech and Hungarian telecom administrations'. Paper presented to 12th Annual NISPAcee Conference, Vilnius May 13-15, 2004;
Morgan, E. and Perry J., (1988) "Re-orienting the Comparative Study of Civil Service Systems." *Review of Public Personnel Administration* , no.3: pp.84-95;
National Training Foundation (NTF), (1998) (1) 'Analysis of Public Administration in the Czech Republic', (2) 'Proposals for a Strategy of Public Administration Reform in the Czech Republic'. Prague: PHARE project;
Nachimias, D. and Nachimias, C., (1981) 'Research Methods in the Social Sciences'. New York: St. Martin's Press;
Nikolaev I., (2005) 'Long Transfer to the Market', in *Apology* N2 2005, Moscow, pp129 – 153;
Nikolaev I., (ed) (2006) 'Strategic Analysis', (in Russian) FBK, Moscow: Delo;
Nozdrachev, A., (1999) 'Gosudarstvennaya sluzhba' (in Russian), Moscow: Statut;
Niskanen, W. A. (1971) 'Bureaucracy and Representative Government', Chicago: Aldine-Atherton;
Nispen tot Pannerden, F.K.M. van & Ringeling, A.B. (2006) The Concept of Subsidiarity in a European Context: The End of National Sovereignty? In D. Eaton (Ed.), *The End of Sovereignty? A Transatlantic Perspective* (Transatlantic Public Policy Series, 2). M: Lit Verlag;
Nispen tot Pannerden, F.K.M. van & Posseth, J.J.A. (2007). Democratic Legitimacy of Economic Governance: The Case of the European and Monetary Union. In V.J.J.M. Bekkers, A.R. Edwards & A.G. Dijkstra (Eds.), *Governance and the Democratic Deficit.* Aldershot: Ashgate;
Nolan C. B., (ed.) (2001) 'Public Sector Reform. An International Perspective'. New York: Palgrave Publishers Ltd;
Nunberg B., (1995) 'Toward a new civil service in Russia: Current issues and future prospects' in *Public Administration Reform in Transition*, Seminar in Salzburg, Austria (November 28 – 1 December 1, 1995);
Nunberg, B., (1998) 'The State After Communism: Administrative Transitions in Central and Eastern Europe', Washington D.C: the World Bank;
Nunberg, B., (2000) 'Ready for Europe. Public Administration Reform and European Union Accession in Central and Eastern Europe.' World Bank Technical Paper N. 466. Washington D.C, the World Bank;
O'Dwyer C., (2001) "Runaway State Building: The Impact of Party System Development On Public Administration Reform in Poland, Slovakia and the Czech Republic, 1991-2000", University of California;
O'Dwyer, C., (2002) 'Civilizing the State Bureaucracy: The Unfulfilled Promise of Public Administration Reform in Poland, Slovakia and the Czech Republic (1990-2000)', Berkeley: University of California;

Obolonsky A., (ed) (1999) 'In the Service of the State: Bureaucracy in New and Old Russia'(in Russian), Moscow: Delo;
OECD, (2001) 'Government Citizens Relations, Country Profile - The Czech Repubic', Paris: OECD Publishing;
OECD, (2001) 'Regulatory reform in the Czech Republic', Paris: OECD Publishing;
OECD, (2003) 'OECD Economic Surveys. Czech Republic' - Volume 2003 Issue 6, Paris: OECD Publishing;
OECD, (2006) 'OECD Economic Surveys. Russian Federation.' Volume 2006, Issue 17, Paris: OECD Publishing;
Offe, C., (1996) 'Designing Institutions in East European Transitions', in R.Goodin (ed.) *TheTheory of Institutional Design,* Cambridge: Cambridge University Press;
Onisko J., (2000) 'Reforme de la Fonction Public d'Etat en Republique Tcheque', Paris: Ecole Nationale d'Administration;
Open Society Institute, (2002) 'Corruption and Anti-Corruption Policy in the Czech Republic', www.eumap.org/reports/2002/corruption/international/sections/czech/;
Otechestvennyie Zapiski, (2004) "Administrative Reform", (in Russian), Moscow, Strana OZ;
Ovsyanko D., (1996) "Civil Service in the Russian Federation" (in Russian), Moscow, Yurist;
Page, B., (1973) 'The Czechoslovak Reform Movement, 1963-1968. A Study in the Theory of Socialism'. Amsterdam: B.R.Gruner B.V.;
Pavlik P., (2002) 'Europeanisation and Transformation of Public Administration: The Case of the Czech Republic', Berlin: Institut für Europäische Politik;
Pehe J., (1996) 'Elections Result in Surprise Stalemate', in *Transition*, N 28, Praha;
Peters, B. G. & van Nispen, F. (2001) Public Policy Instruments: Evaluating the Tools of Public. Administration. Cheltenham: Edward Edgar;
Peters, B.G., (1988) 'Comparing Public Bureaucracies: Problems of Theory and Methods'. Tuscaloosa-London: University of Alabama Press;
Peters, B.G. (1996) 'Theory and Methodology' in Bekke H., Perry J. and Toonen Th. (eds), *Civil Service Systems in a Comparative Perspective*, Bloomington: Indiana University Press, pp. 13-41;
Phare Projekt, (2000) "Modernization of Central State Administration in the Czech Republic", CZ 00 09.01 Inception Report, http://aplikace.mvcr.cz/archiv2008/reforma/phare/ph_2000.html ;
Pierre, J. and Peters B.G., (2003) 'Handbook of Public Administration'. London: Sage.
Pierre, J., (2002) 'Bureaucracy in the Modern State. An Introduction to Comparative Public Administration', Cheltenham, UK: Edward Elgar;
Pollitt C. and Bouckaert G., (2004) 'Public Management Reform. A Comparative Analysis', New York: Oxford University Press;
Potucek M., (1999) 'And what about the state?', *The New Presence*, April 1999, Prague;
Potucek, M., (2003) 'Traveling in the Czech Future and Others', Prague: CESES;
Rauch, J. and Evans P., (1997) "Bureaucratic Structure and Bureaucratic Performance in Less Developed Countries", University of California-San Diego and University of California-Berkeley, manuscript;

Raadschelders Jos C.N. and Rutgers Mark R., (1996) 'The Evolution of Civil Service Systems' in Bekke H., Perry J. and Toonen Th. (eds), *Civil Service Systems in a Comparative Perspective*, Bloomington: Indiana University Press, pp. 67-99;

Raadschelders Jos C.N. (1998) 'Evolution, institutional analysis and path dependency: an administrative-history perspective on fashionable approaches and concepts' in *International Review of Administrative Sciences,* 1998; 64: pp. 565-582;

Raadschelders, Jos C. N. (1998) 'Handbook of Administrative History', New Brunswick: Transaction Books;

Raadschelders, Jos C.N. and Rutgers Mark R. (2001) 'An Overview and Synthesis of Developments and Trends in the Study of Public Administration', Paper presented at the annual conference of the Public Administration Theory Network, 21-23 June 2001, University of Leiden;

Raadschelders, Jos C. N., (2003) 'Administrative History: Introduction' in B.Guy Peters and Jon Pierre (eds) *"Handbook of Public Administration"*. London: Sage Publications, pp. 161 – 167;

Raadschelders, Jos C. N., (2007) 'Tradition in the Study and Practice of Public Administration: Explorations and the Case of Negotiable Authority', Paper presented at the conference "Governing by Looking Back", Australia National University, Canberra, December 12-14, 2007;

Raadschelders Jos C.N., Meer Frits M van der and Toonen Theo A.J. (eds.) (2007) 'The Civil Service in the 21st Century: Comparative Perspectives' Houndsmills: Palgrave/McMillan;

RACS (Russian Academy of Civil Service), (2001) 'The Civil Service of Russia: Problems of Formation and Development', (in Russian), Moscow: RACS;

RACS, (Russian Academy of Civil Service), (2002) "Sociology of power" N 2, Moscow: RACS;

RACS (Russian Academy of Civil Service), (2002) 'Russian Civil Service: Status, Problems and Priorities of Development'. Moscow: RACS;

RACS (Russian Academy of Civil Service), (2004) "Sociology of power" N 1, Moscow: RACS;

RACS (Russian Academy of Civil Service), (2004) "Sociology of power" N 5, Moscow: RACS;

RACS (Russian Academy of Civil Service), (2005) "Sociology of power" N 6, Moscow: RACS;

Rakityansky, N., (2002) 'Modernization of Russia and Political Elite', Vlast No. 1, pp 17-26;

Regular Report on the Czech Republic's Progress Towards Accession, (2001) Report SEC (2001) 1746 of 13 November 2001, European Commission, Brussels;

Rice, M., (1992) 'Public Administration in Post-Socialist Eastern Europe', *Public Administration Review* 52,2, pp. 116-125;

Robinson N., (2002) "Russia. A State of Uncertainty", Oxford: Routledge;

Rusňáková Z., (2004) Presentation at the International Conference 'Democracy and Law in Modern State', 10 June 2004 Prague, Verbatim report;

Russian Academy of Sciences, (2005) 'Bureaucracy and Power in New Russia: Attitudes of Citizens and Expert Assessments', Analytical Report. Moscow: Institute of Sociology;

Rutgers, M. (1998) 'Paradigm lost: crisis of identity of the study of public administration'. *International Review of Administrative Sciences*. Vol. 64/4, p.553-564;

Rutgers, M., (2004), "Comparative Public Administration: Navigating Scylla and Charybdis – Global Comparison as a Translation Problem" *Administrative Theory and Practice* Vol. 26, N 2, June 2004, pp 151 – 168;

Rutgers, M., (ed). (2003) 'Retracing Public Administration', Jay Press/Elsevier International;

Savchenko S., (ed.) (2006) 'Reform of the Russian Civil Service', (in Russian) Moscow: HSE;

Sajo, A., (1998) 'Corruption, Clientelism and the Future of the Constitutional State in Eastern Europe', *East European Constitutional Review,* Vol 7, No 2;

SIGMA (2002) 'Czech Republic. Public Service and the Administrative Framework Assessment 2002', OECD/EC;

SIGMA (2003) 'Czech Republic. Public Service and the Administrative Framework Assessment 2003', OECD/EC;

Solnick, S., (1999) 'Russia's "Transition": Is Democracy Delayed Democracy Denied?' *Social Research*, vol. 66, № 3;

Shablovsky G. and Derlien H., (1993) 'East European Transitions, Elites, Bureaucracies and the European Community', *Governance* Vol. 6, N.3 July;

Sharov S, (2007) Speech at a meeting in High School of Economics on 12 February 2007 of S.Sharov, Head of Department in the Ministry of Economic Development and Trade;

Scherpereel, J., (2004) 'Renewing the Socialist Past or Moving toward the European Administrative Space?' *Administration & Society.*2004; 36: pp 553-593;

Shepherd, R., (2000) 'Czechoslovakia, The Velvet Revolution and Beyond'. London: Macmillan Press Ltd;

Shumpeter J., (1995) 'Capitalism, Socialism and Democracy'. Moscow: Ekonomika;

Slatinov, V., (2006) "Intermediary Results of the Administrative Reform" (in Russian) in *Public Management in XXI century*: *Traditions and Innovations*, Moscow: MSU, pp.460 – 467;

Smorgunov L. (ed) (2002) 'State Management and Policy' (in Russian), StPetersburg: StPetersburg University;

Stewart D, Sprinthall N. and Kem J., (2002) 'Moral Reasoning in the Context of Reform: A Study of Russian Officials', *Public Administration Review*, May/June 2002, Vol. 62, No.3;

Stivers C., (ed) (2000) 'Democracy, Bureaucracy and the Study of Administration'. Cleveland State University: Westview Press;

Streckova Y., (1997) 'Reform of the Czech Public Administration' in *Pubic Management Forum*, N5 September 1997, SIGMA;

Strmiska, M., (2000) 'The Czech Multipartism in the late 1990s', *Stredoevropske politicke studie* vol 2, n 2;

Synnerstrom, S., (1998) "Professionalism in Public Service Management: the Making of Highly Qualified, Efficient and Effective Public Managers." Paper

presented at the OECD conference: 'National and International Approaches to Improving Integrity and Transparency in Government.';
Toonen, Theo A. J. (1993) 'Analyzing Institutional Change and Administrative Transformation: A Comparative View in Administrative Transitions in Central and Eastern Europe, pp. 151-168;
Toonen, Theo A.J. (1998) "Networks, Management and Institutions: Public Administration as 'Normal Science'," *Public Administration,* Vol. 76;
Toonen, Theo A.J., (2001) 'Public Sector Reform in Comparative Perspective: Creating Open Villages and Reinventing the Politics of Administration', in: J. Pierre, B.Guy Peters, (eds), *Politicians, Bureaucrats, and Administrative Reform: the Changing Balance*, London, Edgar Elgar, 2001;
Toonen Theo A.J. (2004) 'Public Sector Reform in the Knowledge Based Economy (KBE), Background Paper to the Keynote Presentation at the CHER 17th Annual conference 'Public-Private Dynamics in Higher Education: Expectations, Developments, Outcomes', Twente University, 17-19 September 2004, Enschede;
Toonen Theo A.J., (2006) 'Analyzing Institutional Change and Administrative Transformation: a Comparative View' (Reprint Toonen, 1993) in: Eric Oteneyo, Nancy Lind (eds), *Comparative Public Administration: the Essential Readings*, Houghton-Mifflin: pp 371-392;
Toonen Theo A.J. (2007) 'Administrative Reform: Analytics' (reprint Toonen 2003) in: B.Guy Peters, Jon Pierre (eds), *The Handbook of Public Administration – Concise Paperback Edition*, Sage Publications, pp 301-311;
Tychtl P., (1999) "Democracy's Disappointment", Central Europe Review Vol.1, No 16, 11 October;
Vachudova M. (2001) 'Ten Years of Czech Party Politics', *The New Presence*, Summer, 2001;
Verheijen T., (2000) 'Administrative Capacity Development', the Hague: WRR;
Verheijen, T., (ed.) (1999) 'Civil Service Systems in Central and Eastern Europe', Cheltenham: Edward Elgar;
Vidlakova O., (1999) 'Civil Service and State Administration. Country Report: Czech Republic', Prague;
Vidlakova O., (2000) 'Comparative Study of Politico-Administrative Relations in CEEC – the Case of the Czech Republic', University Pardubice;
Vidlakova O., (2006) 'The Reform and Modernization of Central State Administration in the Czech Republic'. Country Report, Prague;
Lowndes V., (1996) 'Varieties of New Institutionalism: A Critical Appraisal'. *Public Administration* 74 (2), pp. 181-197;
Vonkova J., (1996), 'Considerations of a Separate Codification of Civil Servants Legal Relations in the State' in *Problems and Outlines of Public Administration in the Czech Republic*, Prague: Phare;
Weber, M. (1947) 'The Theory of Social and Economic Organization'. transl. by A.M. Henderson and Talcott Parsons, New York: The Free Press;
Westminster s.r.o. (2007) 'State Official Yesterday, Today and Tomorrow', Report on the Results of the Poll www.westminster.cz ;
Wilson P., (2005) 'Ottawa in Bohemia', http://paulwilson.ca/pw_articles/pw_sn_outrata.html ;

Wise, Lois R., (1996) "Internal Labor Markets," in Bekke H., Perry J. and Toonen Th. (eds), *Civil Service Systems in a Comparative Perspective*, Bloomington: Indiana University Press, pp. 100-118;
White Jay and Adams Guy (ed) (1994) ‘Research in Public Administration. Reflections on Theory and Practice’, Sage Publications;
Williams K. (1999) “A Scorecard for Czech Lustration”, *Central Europe Review*, Vol.1, No 19;
World Bank, (1997) World Development Report 1997: ‘The State in a Changing World’, New York: Oxford University Press;
World Bank, (1999) ‘Czech Republic. Public Administration Reform’, World Bank Technical Paper;
World Bank (2002) “Transition. The First Ten Years. Analysis and Lessons for Eastern Europe and the Former Soviet Union.” Washington DC: World Bank;
World Bank (2004) ‘Russian Economic Report, N8, June 2004;
World Bank (2006) ‘Institutional Reform in Russia: Moving from Design to Implementation in a Multi-Level Governance Context Study’ Report N 35576-RU;
World Bank, (2006) ‘EU 8. Administrative Capacity in the New Member States. The Limits of Innovation’. Report N 36930-GLB;
Yuzhakov V., (2005) ‘Administrative Reform Only Begins’ in *Profil* N 25, Moscow;
Zaionchkovsky P., (1978) ‘The Government of the Monarchist Russia in the 19th Century’, (in Russian) Moscow: MSU;
Zaitseva T., (ed) (2003) “History of Reform Attempts” (in Russian) Moscow: Ves Mir;
Zyrianov P., (1994) ‘History of Russia’, (in Russian) Moscow: MSU.

Summary

Unprecedented changes in the political and economic systems of the Central and Eastern European (CEE) countries in the late eighties and early nineties of the 20th century highlighted the necessity of transformation of their state administrations and overall systems of governance.

The preparation of civil service reforms in the Russian Federation and the Czech Republic took an extensive period – over a decade after collapse of the communist regimes in Central and Eastern Europe and dissolution of the former Soviet Union and Czechoslovakia. Until the early 2000s the highest authorities in Russia and the Czech Republic effectively ignored civil service reform in terms of its practical implementation. In essence, a kind of contradiction has gradually built up in both Russia and the Czech Republic. The political and economic systems of both countries have drastically changed but the way these countries had previously been administratively managed was largely retained.

The focus of this research is on the identification, analysis and comparison of civil service reform initiatives and processes in the two selected countries – the Russian Federation and the Czech Republic and finding answers to the following major research question: '*What have been the experiences with the preparation, launching and implementation of national civil service reform in the Russian Federation and the Czech Republic following the fall of the communist regimes and which factors can explain the relative success or failure of the reform proposals*'.

The methodology applied in the research is largely based on institutional analysis, which implies "...the notion of systems ruled by norms, culture and traditions rather than by organizational goals and rational perspectives" (Bekke 1999: 4). It thus attaches high importance to the environment in which any institution, including civil service, is embedded. More specifically the methodological approach proceeds from the assumption that existing traditions, values, culture and authority relationships persist through time and exert significant influence on the evolution of the entire civil service system, civil servants' conduct and attitudes, and prospects of civil service reform.

The study draws widely upon the theory, methods and experience of comparative administration accumulated in the works of most prominent scholars and experts in the field. In particular, extensively used were the methodological recommendations contained in the Protocol (guidelines) for Comparative Studies of National Civil Service Systems. The Protocol was designed by academics and experts from Leiden (the Netherlands) and Indiana (USA) Universities and constitutes a common methodological platform (framework) for making meaningful international comparisons of these systems. The research widely applies an inductive method that relies heavily on descriptive data about actual situations and behaviour. It presupposes advance from specific observation to identify differences and commonalities and then progressing to explanations and interpretations.

The choice of the institutional concept as the methodological base of the study determined the overall structure of the research. It incorporates three major interrelated themes: civil service reform capacity; civil service reform implementation; and civil service reform comparison. They are addressed in separate clusters of chapters.

The outcomes of the theoretical and methodological analysis (chapter II) and the inferences made in subsequent country specific chapters (chapters III-VII) allowed identification of factors and forces (variables) that have exerted serious influences on the processes of civil service reform preparation and implementation in the selected countries.

Throughout the 1990s the Russian Federation and the Czech Republic were demonstrating overall low capacity for civil service reform across all of three major parameters:

a) institutional space (framework) - formal and informal institutions (combination/set of rules, norms, traditions and values) prevailing and historical;

b) collective or individual human action (willingness and readiness of the key agents – policy makers and stakeholders to undertake action in the desired direction);

c) resources available (funds and expertise).

The low reform capacity was to a considerable extent predetermined by the historical development and legacy, national and administrative traditions and "path dependency" factors. Widespread patronage and clientilistic networks, excessive politicization of civil services, clear deficit of meritocratic principles, prevailing culture and mentality rooted in the communist and pre-communist past, and insufficient external accountability – all these influences reduced considerably the internal potential for civil service reform in both countries.

Having compared the two cases it was concluded that the capacity for reform had been initially relatively higher in Russia than in the Czech Republic for a number of reasons. Contrary to the situation in the Czech Republic, where reform was driven mainly by external pressure from the European Commission, in Russia the reform's driving force was the President elected in 2000. From the legal point of view the Russian civil service was established in 1993, much earlier than the Czech civil service in 2002. Finally, starting from the early 2000s Russia was no longer experiencing tight budget constraints and could allocate considerably more funds for civil service development and reform.

The identification of key characteristics of the civil service systems in question through the application of the framework research protocol for comparative studies allowed us to conclude that the basic features of merit based civil service systems (appointment and promotion based on open competition, impartiality and performance based remuneration, developed system of job descriptions etc) were largely absent in both Russia and the Czech Republic. In particular we pointed to wide discretionary power of superiors in deciding on civil servants' promotion as well as in determining the size of take home pay of their subordinates.

After lengthy and difficult periods of delay and uncertainty, civil service reforms in the Russian Federation and the Czech Republic were eventually launched in 2003-2004. Out of a whole number of reasons explaining these delays and difficulties two major interrelated factors have served as key constraints in launching civil service reform: a) high level of politicization of both civil services complemented by widespread clientilistic and patronage networks; and b) seriously weakened political competition between main political parties/coalitions both in the Czech Republic and Russia.

Notable factors, differing in each country, played a decisive role in triggering civil service reforms. In the Czech case it was permanent pressure applied by the European Commission in view of the planned accession of the Czech

Republic to the European Union. In the Russian case it was the election of a new President who committed himself to reform the state bureaucracy.

The civil service reform agendas in the two countries were formulated mainly on the basis of information derived from the study of modern trends in administrative reforms and other countries' experiences in this field. Overall the reform agendas appear to be mixed containing priorities adhering to traditions of both *governance* (civil service is primarily a mechanism for effective governability and political legitimacy) and *management* (attention above all to efficiency and economic rationality). However, in practical implementation of reform in both countries distinct priorities were given to the measures directed at strengthening aspects of efficiency of civil services. Issues of legitimacy, accountability and transparency in their practical enforcement turned out to be quite "unpopular" among bureaucratic officialdom in both Russia and the Czech Republic.

The following features of civil service reforms in Russia and the Czech Republic determined largely their current level of advance and results:

- the reforms were designed and prepared in a confidential non-transparent manner (no real involvement of citizens, business, non-government organisations and civil servants themselves);
- they have been prepared and implemented in a top-down fashion (orders and instructions) leaving little or no place for "grass root" initiative;
- there has been no real ownership of reform and this complemented by a dearth of committed stakeholders;
- as a rule, criteria of reform effectiveness and achievement of its results have been either vague and general or totally absent leading to a situation when matching them against practice was extremely difficult;
- civil society and the business community were not invited to monitor the advancement and progress of reform;
- operational management of the reforms appears to have been overall weak.

In terms of results of civil service reforms achieved so far in Russia and the Czech Republic, it is necessary to underline the major achievement that the reforms have actually been launched. However, reforms in both cases have been characterized by a wide mismatch between intentions and results so far. Specific improvements in the two civil services and their environment overall have been modest.

Development of reform of the Czech central state administration has so far resulted in: adoption of the Civil service Act; creation of the Institute of Public Administration; elaboration of methodological approaches for introduction of the planned innovations in civil service and HR management; pilot application of the CAF model in central government; expanded programmes of civil servants' post-entry training; and wider application of IT in the Czech ministries.

Major current achievements of the civil service reform in Russia are confined essentially to: the adoption of new civil service legislation; restructuring of the government (as part of administrative reform); development and adoption of modern job descriptions; implementation of pilots in such areas as introduction of performance based pay and standards of services; elaboration of methodological tools for introducing other planned innovations; and increased remuneration of civil servants.

In terms of pace of reform, we can speak about a slow and cautious bordering into a "no movement at all" approach to reform of the central state administration in the Czech Republic (2004 – 2006) and a rather "big bang", "hasty" approach in restructuring the Russian Government in 2004 and relatively more gradual intervention for implementation of the Russian civil service reform programme (2003–2007).

Our findings conclude that the reform measures undertaken in the two countries, though generally appropriate, have not been sufficiently focused and consistent in addressing the problems of civil service development. Further, the reforms in both countries have been running counter to the prevailing bureaucratic and administrative culture and practices and therefore met with "silent" but overall significant resistance. Finally, support to the reform from the political and administrative elite has been so far largely superficial and conditional. The further the reform implementation proceeded, the less convincing this support appeared to be.

Most importantly, in both Russia and the Czech Republic the reforms have failed so far to introduce changes in the internal incentive mechanism of the civil service and its orientation, as well as to establish effective external control over the service. The weakest point of reform management and strategic approach in both cases was excessive and even exclusive reliance on legislation as the main instrument of reform implementation and the absence of permanent and effective management structures enjoying broad support to their mission. This situation was further compounded by the problem of generally weak law enforcement in both countries.

Within the identified set of variables certain factors appear to have played a decisive role in defining the actual direction and pace of reforms as well as determining the results achieved so far. These are excessive politicization of the civil services, wide spread patronage networks, and the administrative culture and leadership factors. A distinct feature of these key variables is that their origin is heavily influenced by historical legacy and path dependence.

In pursuit of answers to the research question and sub-questions we have established that clientilistic and patronage networks (widespread in the former communist countries) have survived political upheavals and remain the basis of relationships across state administrative institutions in the Russian Federation and the Czech Republic. These networks are "reinforced" by the high degree of politicisation of the Russian and the Czech civil services. Political parties and leaders view state administration as an instrument of their personal power. In most instances members of such networks tend to maintain a mutually beneficial status quo, support broad opportunities for discretion and resist any change.

The largely uncompetitive political environment that existed in Russia and the Czech Republic throughout most of the 1990s reduced to a minimum opportunities for "correcting" the policies of "little or no interest at all" in launching reform of civil services. In essence the Czech Prime Minister V. Klaus and the Russian President B. Yeltsin (who both were reluctant to start serious reform of state administration) were under no effective pressure (which would usually come from the political opposition, parliament or civil society) to modernize and reform the existing system of state administration.

By matching the two civil service reforms and the impact of the same set of variables we were able to define commonalities and distinctions in their contents, driving forces, rate of advance and outcomes. Similarities in reform preparation and implementation can be in particular attributed to the largely common roots of administrative culture and mentality which can be traced to the heritage of monarchist (Austro-Hungarian and Russian Empires) and recent communist regimes.

In spite of a considerable number of similarities the reforms in the Russian Federation and the Czech Republic were distinguished by a number of significant differences. The variance is manifested above all in the major driving forces of reform (European Commission in the Czech case and new President in the Russian case) and in the status of reforms by 2008 (continuation of civil service reform in Russia and its suspension in the Czech Republic).

The causes of major differences between the two countries in the civil service reform processes and current outcomes largely stem from the distinctions in the existing institutional configurations of the two states. As an example, one can refer to the peculiarities of their constitutional framework and political regime: in the Czech Republic - rotating political coalitions concerned to preserve the right to "spoils"; in the Russian Federation - the President with relatively more powers in enforcing his policies and much less dependent on the narrow interests of political parties.

Our study has also reaffirmed our assertion that no matter how radical and innovative declarations of reforms in Russia and the Czech Republic could be (under the influence of either internal or external factors), the extent of their implementation may deviate considerably from proclaimed intentions. In particular we realized that the political and bureaucratic elites in the two countries had largely simulated and distorted the essence of civil service reform that could limit their discretion. There has been strong tendency to confine civil service reform to purely legal and organizational (technical) issues and to make adoption of civil service law the starting and finishing point of the reform.

By 2007 civil service reform in the Czech Republic had been virtually put on hold. This fact was largely pre-determined by a combination of factors including: overall initial low capacity for civil service reform; inertia and resistance of political and bureaucratic elites fearing that upcoming changes would undermine their privileged position and existing clientilistic networks (right for "spoils"); administrative traditions and culture rooted in the communist and pre-communist past; failure to win broad support for reform from civil servants, businesses and civil society; and virtual cessation of the pressure coming from the European Commission following accession of the Czech Republic to the European Union in 2004.

By 2007 the civil service reform in Russia appeared to be advancing further albeit at a slower rate with a number of principal outcomes and outputs still outstanding. In spite of certain setbacks in its implementation the civil service reform remains an important issue of the government agenda in Russia. The prospects for success of civil service reform in Russia depend to a considerable extent on the continuity of the reform policies after completion of the President Putin's term in office in 2008. Should the reform be suspended or fail to reach its major objectives, the current trends in the country's political and economic development may lead to eventual 'privatization of the Russian state by its bureaucracy'.

Samenvatting

De zeer omvangrijke veranderingen in de politieke en economische systemen van Centraal- en Oost-Europese (COE) landen, die later in de jaren '80 en vroeg in de jaren '90 van de twintigste eeuw zin opgetreden, vormen een uitdrukking van de toen gepercipieerde urgentie om de respectievelijke staatsbesturen en het daarbij behorende bestuurssystemen te hervormen.

De aanloop van de ambtelijke hervormingen in de Russische Federatie en de Tsjechische Republiek heeft een aanzienlijke tijd in beslag genomen; langer dan een decennium na het ineenstorten van de communistische regimes in Centraal- en Oost-Europa en het uiteenvallen van de voormalige Sovjet Unie en Tsjecho-Slowakije. Tot in het begin van het eerste decennium van dit millennium hebben de hoogste gezagsdragers in Rusland en Tsjechië ambtelijke hervormingen in termen van een daadwerkelijke implementatie genegeerd. In wezen is geleidelijk aan zowel in Rusland als Tsjechië een contradictie binnen de veranderingsprocessen ontstaan. De politieke en economische systemen van beide landen zijn drastisch veranderd, maar de manier waarop deze landen worden bestuurd, is voor een groot deel gelijk gebleven.

De focus van dit onderzoek is gericht op de vaststelling, de analyse en de vergelijking van initiatieven en processen in beide landen waar het de hervorming van het ambtelijke systeem betreft. Daarbij wordt een antwoord gezocht op de volgende onderzoeksvraag: *'Wat zijn de ervaringen met de voorbereiding, de start en de implementatie van hervormingen binnen de nationale ambtelijke dienst in de Russische Federatie en de Tsjechische Republiek na de val van de communistische regimes en welke factoren kunnen het relatieve succes of falen van de hervormingsvoorstellen verklaren?'.*

De aanpak gebruikt in dit onderzoek, sluit aan bij de neo-institutionele benaderingswijzes. Dit impliceert de veronderstelling "...dat systemen eerder worden beheerst door normen, cultuur en tradities dan door doelen van organisaties en rationele perspectieven" (Bekke 1999:4). Belang wordt hecht aan de omgeving, waarin een institutie, inclusief het ambtelijk apparaat, is ingebed. Meer specifiek gaat deze benadering uit van de veronderstelling dat bestaande tradities, waarden, cultuur en autoritaire relaties grotendeels overeind blijven staan bij het verstrijken van tijd en een belangrijke invloed kunnen uitoefenen op de ontwikkeling van het ambtelijk systeem, het gedrag en de opvattingen van ambtenaren en de realisering van ambtelijke hervormingen.

In dit onderzoek wordt gebruik gemaakt van de theoretische en methodische inzichten van de vergelijkende bestuurskunde. In het bijzonder wordt benadering neer gelegd in het "Protocol for Comparative studies of national civil services"in deze studie gevolgd. Dit protocol is geformuleerd door onderzoekers van de universiteiten van Leiden (Nederland) en Indiana (VS) en vormt een gemeenschappelijk methodologisch kader om systematische internationale vergelijkingen tussen deze ambtelijke systemen mogelijk te maken.

De keuze voor de institutionele theorie als basis van de studie heeft de structuur van dit onderzoek in sterke mate bepaald. Er zijn drie onderling verbonden thema's vastgesteld: de capaciteit tot ambtelijke hervorming, de implementatie van hervorming van de ambtelijke dienst; en vergelijking van ambtelijke hervormingen. Deze zijn behandeld in aparte clusters van hoofdstukken.

De uitkomsten van de theoretische analyse (hoofdstuk II) en de conclusies getrokken in de daarop volgende landspecifieke hoofdstukken (hoofdstukken III-VII) hebben geleid tot de vaststelling van factoren, die belangrijke invloed hebben uitgeoefend op de voorbereiding op en implementatie van hervormingen binnen de publieke sector van de Russische Federatie en de Tsjechische Republiek. Gedurende de jaren '90 toonden de Russische Federatie en de Tsjechische Republiek een geringe capaciteit ambtelijke hervormingen op drie parameters te realiseren te weten:

a) institutionele ruimte (kader) – formele en informele instituties (combinatie/set van regels, normen, tradities en waarden) bestaand in heden en verleden;
b) collectieve of individuele menselijke actie (wil en bereidheid van sleutel actoren – beleidsmakers en belanghebbenden om actie te ondernemen in de gewilde richting);
c) beschikbare middelen (fondsen en expertise)

De geringe hervormingscapaciteit wordt voor een belangrijk deel bepaald door de eigen historische ontwikkeling en erfenis, de nationale en bestuurlijke tradities en "pad afhankelijke" factoren. Wijdverspreide patronage en cliëntelistische netwerken, een sterke mate van politisering van het ambtelijk apparaat, een gebrek aan meritocratische principes, de heersende cultuur en mentaliteit geworteld in het communistische en precommunistische verleden, onvoldoende externe verantwoordelijkheid—dit alles verminderde het potentieel tot ambtelijke hervorming in beide landen aanzienlijk.

Nadat de twee cases zijn vergeleken, is geconcludeerd dat de hervormingscapaciteit in Rusland bij aanvang relatief hoger is geweest dan in Tsjechië. Een aantal redenen is daarbij van belang geweest. In tegenstelling tot de situatie in Tsjechië, waar de hervormingen voornamelijk gedreven werd door druk van de Europese Commissie, was in Rusland de in 2000 verkozen president de drijvende kracht. Vanuit een juridisch oogpunt is in Rusland de ambtelijke dienst (1993), eerder ontstaan dan de Tsjechische ambtelijke dienst (2002). Vanaf het begin van de 21^{e} eeuw heeft Rusland niet meer te maken gehad met de beperking van een krap budget. Men kon aanzienlijk meer middelen voor ontwikkeling en hervorming van de ambtelijke dienst toewijzen.

De bepaling van de basiskenmerken van de respectievelijke ambtelijke systemen door middel van de toepassing van het eerder genoemde onderzoeksprotocol heeft als conclusie opgeleverd dat criteria als benoeming en promotie gebaseerd op open competitie, onpartijdigheid en beloning op basis van prestaties, een ontwikkeld systeem van baanbeschrijvingen etc.) grotendeels zowel in Rusland als Tsjechië afwezig zijn geweest. In het bijzonder is gewezen op de ruime mate van beleidsvrijheid van superieuren bij hert nemen van beslissingen over de promotie van ambtenaren en de bepaling van het loon dat onderschikten mee naar huis mochten nemen.

Na een lange en moeizame periode van vertraging en onzekerheid, zijn de ambtelijke hervormingen in Rusland en Tsjechië uiteindelijk gestart in de jaren 2003 en 2004. Van het aanzienlijke aantal factoren dat verantwoordelijk voor deze vertraging kan worden gesteld, zijn er twee gemeenschappelijke factoren in het bijzonder als belangrijkste obstakels te beschouwen: a) een grote mate van politisering van beide ambtelijke diensten gecombineerd met wijdverspreide cliëntelistische en patronage netwerken; en b) geringe mate van politieke competitie tussen belangrijke politieke partijen/coalities. Er zijn ook verschillen tussen beide landen waar te nemen. In de Tsjechische casus was er een permanente druk van de Europese Commissie aanwezig gegeven de geplande toetreding van Tsjechië tot de Europese Unie. In de Russische casus is de verkiezing van een nieuwe president belangrijk geweest, die zichzelf heeft gecommitteerd aan de hervorming van de nationale bureaucratie.

De agenda's van de ambtelijke hervormingen in de twee landen zijn vooral geformuleerd op grond van onderzoek naar nieuwe trends in bestuurlijke hervormingen en aan de hand van ervaringen op dit gebied van andere landen. Over het algemeen blijken de hervormingsagenda's te bestaan uit een mengsel van prioriteiten behorend tot zowel de *governance* (het ambtelijk apparaat wordt voornamelijk gezien als een instrument voor effectieve bestuurbaarheid en politieke legitimiteit) en *management* (er is vooral aandacht voor efficiëntie en economische rationaliteit) tradities. Echter, bij de feitelijke implementatie van hervormingen in beide landen is prioriteit gegeven aan de maatregelen die gericht waren op een versterking van de efficiëntie van het ambtelijk apparaat. Onderwerpen als legitimiteit, aansprakelijkheid en transparantie kunnen zowel binnen bureaucratische kringen in Rusland als Tsjechië op weinig populariteit rekenen.

De volgende kenmerken hebben in grote mate de (geringe) mate voortgang en succes van de hervormingen bepaald:

- de hervormingen zijn ontworpen en voorbereid in een vertrouwelijke non-transparante manier (geen werkelijke betrokkenheid van burgers, bedrijfsleven, niet-overheidsorganisaties en ambtenaren zelf);
- ze zijn op een top-down wijze voorbereid en geïmplementeerd zonder veel ruimte te laten voor initiatieven van onderaf;
- de belanghebbenden hebben de hervorming niet gezien als hun eigendom, waardoor er geen steun onder de betrokkenen is gegenereerd;
- als regel zijn de criteria voor effectiviteit van hervorming en het behalen van resultaten ofwel vaag en algemeen van aard of zelfs afwezig waardoor de implementatie uitermate moeizaam wordt;
- het maatschappelijk middenveld en het bedrijfsleven zijn niet uitgenodigd om te participeren in de bepaling en monitoring van de voortgang van de hervormingen;
- het operationeel management van de hervormingen blijkt in het algemeen zwak ontwikkeld te zijn geweest.

In termen van resultaten van ambtelijke hervormingen die tot op heden zijn bereikt in Rusland en Tsjechië, is het nodig het belangrijkste succes, dat wil zeggen dat er hervormingen van de ambtelijke dienst zijn gestart, te onderstrepen. Echter, er bestaat een tegenstelling tussen de intenties en de bereikte resultaten. Specifieke verbeteringen in de beide ambtelijke apparaten zijn tot dusverre bescheiden te noemen.

In het kader van de voorbereiding en implementatie van hervormingen in het Tsjechische centrale staatsbestuur hebben de bereikte resultaten zich voornamelijk geuit in de vorm van: het aannemen van een ambtenarenwet; het creëren van een

Instituut voor het Openbaar Bestuur; het uitwerken van methodologische benaderingen voor de introductie van voorgenomen verbeteringen van de ambtelijke dienst en Human Resource Management initiatieven; een proeftoepassing van het CAF-model binnen de centrale overheid; uitgebreidere trainingsprogramma's van ambtenaren; en bredere toepassing van IT in Tsjechische ministeries.

Resultaten tot nu toe in de hervormingen van het ambtelijk systeem in Rusland zijn beperkt gebleven tot: het aannemen van wetgeving voor de ambtelijke dienst; een herstructurering van de overheid (als onderdeel van bestuurlijke hervormingen); de ontwikkeling en opnemen van moderne functiebeschrijvingen; de implementatie van pilots als de introductie van prestatiegericht belonen en standaarden voor dienstverlening; de uitwerking van een instrumentarium om andere voorgenomen innovaties te introduceren en een hogere beloning voor ambtenaren.

In termen van de snelheid van hervormingen, kunnen we spreken van een langzame en voorzichtige (grenzend aan een "helemaal geen beweging") benadering in het geval van de hervorming van het centrale staatsapparaat in Tsjechië (2004-2006) en een "big bang" en bijna "gehaaste' benadering van de Russische overheid om het ambtelijk systeem te herstructureren in 2004 en een relatief meer geleidelijke implementatie van het Russische hervormingsprogramma van de ambtelijke dienst daarna.

Deze bevindingen leiden tot de conclusie dat de genomen hervormingsmaatregelen in de twee onderzoekslanden, hoewel over het algemeen passend, niet genoeg gefocust en consistent zijn geweest om de problemen van de ambtelijke dienst aan te pakken. Daarnaast staan de hervormingen in beide landen haaks op de bestaande bureaucratische en bestuurlijke cultuur en praktijk. Ze hadden daarom te maken met "stille" maar overal merkbare weerstand. Tenslotte, is de steun voor de hervorming van de politieke en bestuurlijke elite tot op heden voornamelijk oppervlakkig en conditioneel van aard gebleken. Hoe verder de implementatie van de hervorming vorderde, hoe minder overtuigend de steun lijkt te zijn geweest.

Zowel in Rusland als Tsjechië hebben de hervormingen tot dusver geen daadwerkelijke veranderingen tot stand gebracht in de intrinsieke motivatiemechanismen binnen de ambtelijke dienst. Evenmin is er een effectieve externe controle over het ambtelijk apparaat gerealiseerd. Het zwakste punt van het management van de hervormingsprogramma's is de overmatige en het bijna exclusieve vertrouwen in wetgeving als belangrijkste instrument voor de implementatie van hervormingen geweest en de afwezigheid van permanente en effectieve managementstructuren gericht op het genereren van een brede steun voor de projecten Deze situatie wordt nog verder gecompliceerd door het probleem van zwakke wetshandhaving in beide landen.

Binnen de in het onderzoek benoemde verklarende variabelen blijken de volgende factoren een beslissende rol te hebben gespeeld bij de richting en snelheid van de hervormingen en de behaalde resultaten. Deze cruciale factoren zijn de sterke mate van politisering van het ambtelijk apparaat, de wijdverspreide patronagenetwerken, en de vigerende bestuurscultuur en het aanwezige leiderschap. Deze factoren hebben een worteling in de nationale historie en bepalen daarmee de padafhankelijkheid van ontwikkelingen. In speurtocht naar antwoorden op de onderzoeksvraag en deelvragen hebben we gezien dat de cliëntelistische- en patronagenetwerken de politieke omwentelingen hebben overleefd en het fundament zijn blijven vormen van de bestuurlijke staatsinstituties zowel in Rusland als Tsjechië. Deze netwerken worden "versterkt" door een hoge graad van politisering van de Russische en Tsjechische ambtelijke diensten. Politieke partijen en leiders zien het staatsbestuur als een instrument van hun persoonlijke macht. In de meeste gevallen willen de leden van zulke netwerken de gunstige status quo behouden en daarmee hun ruime eigen beleidsvrijheid. Vandaar verzet men zich tegen voorgenomen veranderingen.

De geringe mate van politieke concurrentie in zowel Rusland als Tsjechië gedurende het grootste deel van de jaren '90 heeft de mogelijkheden om de desinteresse in het hervormingsbeleid te verminderen. In essentie stonden de Tsjechische Minister-President V. Klaus en de Russische President B. Yeltsin - die beiden terughoudend waren bij het entameren van een serieuze hervorming van het staatsbestuur - niet onder daadwerkelijke druk van de politieke oppositie, het parlement of het maatschappelijk middenveld om het bestaande systeem van staatsbestuur te moderniseren en hervormen.

Door de ambtelijke hervormingen en de effecten van de zelfde set variabelen in beide landen te vergelijken, is het mogelijk geweest de overeenkomsten en verschillen te bepalen voor wat betreft de inhoud, stuwende krachten, vooruitgang en uitkomsten. Overeenkomsten in hervormingsvoorbereiding en –implementatie kunnen in het bijzonder worden toegeschreven aan de grotendeels overeenkomstige bestuurscultuur en mentaliteit die teruggevoerd kan worden op de erfenis van monarchistische (het Oostenrijks-Hongaarse Rijk en het Tsaristische Rusland) en recente communistische regimes.

Naast een aanzienlijk aantal overeenkomsten zijn er ook een aantal belangrijke verschillen tussen de hervormingen in Rusland en Tsjechië te constateren. De variatie doet zich vooral voor in de aard van de belangrijkste stuwende krachten achter de hervorming (de Europese Commissie in de Tsjechische casus en de nieuwe president in de Russische casus) en daarnaast in de recente status (2008) van de hervormingen; namelijk een continuering van ambtelijke hervormingen in Rusland en de vertraging daarvan in Tsjechië.

De verklaring van de verschillen tussen beide landen voor wat betreft de processen en uitkomsten van de hervormingen in het ambtelijk systeem zijn gelegen in de onderscheiden institutionele configuraties. Bij wijze van voorbeeld kan gerefereerd worden aan de kenmerken van het vigerende constitutionele kader en het politieke regime. In Tsjechië kan gewezen worden op de wisselende politieke coalities die hun recht op "spoil" benoemingen wensen te behouden en in Rusland op de aanwezigheid van een president die beschikt over relatief meer machtsmiddelen om zijn beleid af te dwingen en daarom minder afhankelijk is van de belangen van politieke partijen.

Deze studie heeft aangetoond dat, hoe radicaal en innovatief hervormingsintenties in Rusland en Tsjechië ook kunnen zijn, de mate van hun implementatie substantieel kan afwijken van de voorgenomen plannen. In het bijzonder is naar voren gekomen dat de politieke en bureaucratische elites in beide landen de kern van de ambtelijke hervormingenvoorstellen, die hun beleidsvrijheid zouden hebben kunnen beperken, voor een groot deel hebben tegengehouden. Er bestaat een tendens om hervormingen van het ambtelijk apparaat te beperken tot puur juridische en organisatorische (technische) aspecten en het aannemen van de ambtenarenwet te beschouwen als start- en eindpunt van de hervormingen.

Tegen 2007 is de ambtelijke hervorming in Tsjechië praktisch tot stilstand gekomen. Dit wordt veroorzaakt door een combinatie van factoren zoals: een lage initiële hervormingscapaciteit; de traagheid van en weerstand aanwezig bij politieke en bureaucratische elites die vrezen dat toekomstige veranderingen hun bevoorrechte positie en bestaande cliëntelistische netwerken (en hun recht op "spoils") zouden ondermijnen; bestuurlijke tradities en cultuur geworteld in het communistische en precommunistische verleden; het falen om brede steun voor hervorming te winnen bij ambtenaren, bedrijven en het maatschappelijk middenveld en de verminderde druk van de Europese Commissie volgend op de toetreding van Tsjechië tot de Europese Unie in 2004.

In 2007 lijkt de ambtelijke hervorming in Rusland verder te gaan. Ook al gebeurt het langzamer, er liggen een aantal resultaten in het vooruitzicht. Ondanks enkele tegenvallers in de uitvoering blijft de hervorming van de ambtelijke dienst een belangrijke plaats op de overheidsagenda in Rusland innemen. De vooruitzichten voor succes van de ambtelijke hervorming in Rusland hangt voor een aanzienlijk deel af van de continuïteit van het hervormingsbeleid nu de termijn van president Poetin in 2008 versteken is. Als de hervorming faalt, waar dit het behalen van de belangrijkste doelen betreft, dan kunnen huidige trends in de politieke en economische ontwikkeling van het land eventueel leiden tot 'privatisering van de Russische staat door zijn bureaucratie'.

CURRICULUM VITAE

Alexander Kotchegura was born in Magdeburg in Germany on 3 June 1956. He is a University Lecturer and Public Management Consultant with over ten years record of lecturing to high school students and civil servants in two leading Moscow Universities and over fifteen years experience of work as an expert and manager in various projects funded by the World Bank, European Commission, UNDP, British and Canadian governments. His expertise builds upon lengthy record of practical civil service, project management and academic employment. He is an Editorial Board member of the "International Public Management Journal" (ISSN: 1096-7494) and has publications in Russia, Great Britain and the United States. Currently he holds the post of Deputy Head of Chair of State and Municipal Management at the Faculty of Humanities and Social Sciences of the Russian Peoples' Friendship University. In 1996 Alexander Kotchegura joined an international consortium of scholars (led by academics from Dutch and American universities) which launched comparative research of civil service systems across various countries.

Decree of the President of the Russian Federation
No. 1336 of November 19, 2002

ON FEDERAL PROGRAM "REFORM OF THE CIVIL SERVICE OF THE RUSSIAN FEDERATION (2003-2005)"[1]

For the purpose of improving the efficiency of the civil service in order to strengthen the state and develop the civic society, I hereby decree as follows:

1. To consider it necessary to implement in 2003-2005 a set of measures aimed at improving legal, organizational, financial, and methodological framework of the civil service of the Russian Federation.

2. To establish the following main areas of reform of the civil service of the Russian Federation:

- Creating a comprehensive normative and legal framework governing the civil service of the Russian Federation;
- Developing efficient mechanisms for implementing a personnel policy with regard to the civil service of the Russian Federation for the purpose of optimizing the structure of civil service personnel;
- Developing measures aimed at improving compensation of civil servants and financial, economic, and logistic support of the civil service of the Russian Federation, and at using advanced information technology in the civil service system in a rational manner;
- Implementing staff training programs for the civil service of the Russian Federation and professional development programs for public servants;
- Forming a system of administration of the civil service of the Russian Federation.

3. To approve the attached Federal Program "Reform of the Civil Service of the Russian Federation (2003 – 2005)" (hereinafter referred to as the Program).

4. To appoint the Administration of the President of the Russian Federation as the Program Coordinator.

5. The Government of the Russian Federation shall provide funding for the Program when developing draft federal budgets through the year 2005.

6. The Chief of the Administration of the President of the Russian Federation shall:

---

[1] Source: Rossiyskaya Gazeta No. 223 of November 23, 2002

Designate a person responsible for implementing the measures envisaged by the Program;
Update the President of the Russian Federation on an annual basis on the progress in implementing the measures envisaged by the Program.

7. To advise the state bodies of the constituent entities of the Russian Federation, within the limits of their budget funds:

To participate in implementing the measures envisaged by the Program;
To develop and approve programs of reform of the civil service in the constituent entities of the Russian Federation.

8. This Decree shall enter into force at the date of its official publication.

Vladimir Putin

President of the Russian Federation

FEDERAL PROGRAM "REFORM OF THE CIVIL SERVICE OF THE RUSSIAN FEDERATION (2003-2005)"

Approved by Decree of the President of the Russian Federation No. 1336 of November 19, 2002

Passport of Federal Program "Reform of the Civil Service of the Russian Federation (2003 – 2005)"

Program name – Federal Program "Reform of the Civil Service of the Russian Federation (2003 – 2005)".

Name, number, and date of the decision to develop the Program – Instruction of the President of the Russian Federation No. Pr-1496 of August 15, 2001.

Program Coordinator – Administration of the President of the Russian Federation.

State customers of the Program – Ministry of Labor and Social Development of the Russian Federation, Ministry of Trade and Economic Development of the Russian Federation, Ministry of Finance of the Russian Federation, Ministry of Internal Affairs of the Russian Federation, and Ministry of Defense of the Russian Federation.

Principal Program implementers – Ministry of Labor and Social Development of the Russian Federation, Ministry of Trade and Economic Development of the Russian Federation, Ministry of Internal Affairs of the Russian Federation, Ministry of Defense of the Russian Federation, Ministry of Property Relations of the Russian Federation, Ministry of Finance of the Russian Federation, Ministry of Justice of the Russian Federation, State Committee for Construction and Housing of the Russian

Federation, research and development, educational, and other entities to be determined on a tender basis according to the established procedure.

Program goal – improving the efficiency of the civil service as a whole and of its types and levels, optimizing civil servant maintenance costs, and developing civil service resources.

Program objectives –

- Creating conditions for optimal organizational and legal support of the civil service;
- Defining roles, powers, and responsibilities of civil servants on the basis of job (service) descriptions;
- Implementing new techniques for planning, financing, stimulating, and assessing civil servants' activities, and using civil service system resources in a rational manner;
- Ensuring openness of the civil service for the benefit of civic society development and strengthening of the state;
- Applying efficient methods for selecting qualified personnel for the civil service and for assessing professional performance of civil servants, as well as creating conditions for their job (service) promotion;
- Implementing staff training programs for the civil service and professional development programs for civil servants;
- Implementing mechanisms for identifying and solving public service-related conflicts of interest, as well as introducing a legal regulation of professional ethics of civil servants;
- Creating an optimal material and technical environment for efficient functioning of the federal civil service and for performance by civil servants of their official (service) duties;
- Ensuring the development of a civil service management system.

Program time-frame – 2003–2005

Principal measures envisaged by the Program –

- Improving the legislative and regulatory framework for civil service reform issues;
- Experimenting and implementing pilot projects aimed at applying new approaches to federal civil service organization and support of federal public servants' activities;
- Improving training, retraining, and professional development of federal civil servants;
- Creating a material and technical environment for efficient functioning of the federal civil service;
- Forming a civil service management system.

Program funding volumes and sources - federal budget funds of RUR 539.1 million for the entire period of the Program implementation, including:

Research and development work – RUR 296.1 million;

Other – RUR 243 million.

Program deliverables –

- Harmonizing the regulatory framework of the civil service with the existing social relations and new economic environment;
- Establishing a professional civil service based on the principles of taking into account and assessing civil servants' performance;
- Implementing job (service) descriptions for civil servants;
- Achieving a high-quality level of performance by civil servants of their official (service) duties and of provision by them of public services to citizens and organizations;
- Creating conditions for openness and accountability of state bodies' administrations and civil servants to the civic society;
- Creating a framework for improving financial and economic support of the federal civil service;
- Improving the efficiency of personnel policy within the civil service system in order to improve the quality of civil service employee structure;
- Implementing mechanisms for identifying and solving public service-related conflicts of interest, as well as introducing a legal regulation of professional ethics of civil servants;
- Raising the professional level of civil servants;
- Optimizing the federal civil servant headcount;
- Creating a material and technical environment for efficient performance by civil servants of their official (service) duties;
- Forming a civil service management system.

Program implementation control system – the Program Coordinator shall be responsible for monitoring Program implementation.

1. Problem description and substantiation of the need to solve it using the methods envisaged by the Program

The civil service's constitutional framework developed in the Russian Federation defines it as a state government mechanism that is thoroughly different from the previous administrative system.

A host of public service-related laws and regulations were developed in 1992–2002. Federal Law On the Fundamentals of the Public Service of the Russian Federation, statutes and other regulations of the Russian Federation governing various types of the federal civil service as well as the civil service of constituent entities of the Russian Federation were enacted.

The total number of civil servants and other employees of federal state bodies, state bodies of the constituent entities of the Russian Federation, and other state bodies established under the Constitution of the Russian Federation (hereinafter referred to as the state bodies), as well as municipal servants and other employees of local self-government bodies at the beginning of 2002 amounted to 1,053.1 thousand, including a verified 315.1 thousand public servants and other employees of federal executive bodies. For the period of 1992 to 2002, the number of civil servants and other employees of these bodies grew by a factor of 1.8, primarily at the regional level. This growth was due to the establishment under the Constitution of the

Russian Federation of state bodies responsible for the development of the market economy, including tax, finance, employment, migration, and other bodies, as well as local self-government bodies.

Civil servants and other employees of executive state bodies, as well as municipal servants and other employees of local self-government bodies (925.1 thousand) accounted for 89 percent of the total number of civil servants and other employees of state bodies, municipal servants, and other employees of local self-government bodies; 30 percent of them were employed in federal executive bodies, including 28 percent employed in territorial bodies of federal executive bodies.

The verified number of federal civil servants of central administrations of federal executive bodies amounted to 24.9 thousand, and of territorial bodies thereof, to 290.2 thousand.

State bodies of the constituent entities of the Russian Federation employed 153.3 thousand civil servants, while local self-government bodies employed 283.7 thousand municipal servants.

A decrease in the number of civil servants of federal executive bodies has been noted since 1998. In 2001, the number of civil servants of federal executive bodies decreased by 4.9 percent compared to 1995. At the same time, the number of civil servants of state bodies of the constituent entities of the Russian Federation and municipal servants of local self-government bodies is on the rise.

In 1999, Russia had 2.8 civil servants of federal executive bodies per 1,000 people of total population, while the UK had 5.2, and the U.S. had 3.9; the ratios per 1,000 people of employed population amounted to 6.4, 11.3, and 7.9, respectively.

In the process of civil service reform, it is envisaged to optimize the number of civil servants by implementing new civil service administration techniques.

The existing gender and age structure of the civil service does not match the gender and age structure of Russia's employed population.

Thus, at the beginning of 2002 women accounted for over 70 percent of all civil servants employed in public positions, including 71.8 percent in executive bodies, 67.6 percent in administrations of judicial bodies and prosecutors' offices, 55.8 percent in administrations of legislative bodies, and 55.4 percent in administrations of other state bodies. The relatively high percentage of women in administrations of state bodies is due to the prevalence of women in public offices classified as senior and junior position groups.

The average age of civil servants employed in public positions was 40 years. Administrations of federal courts and prosecutors' offices of the Russian Federation employed somewhat younger personnel (average age 37).

Persons in retirement age accounted for 3.5 percent of all civil servants. Over 30 percent of civil servants employed in administrations of legislative and other state bodies were 50 years old and older, including 11.1 percent and 6.9 percent respectively of those in retirement age.

Most civil servants aged up to 30 were employed in public offices classified as senior and junior groups of positions (49.6 and 38.7 percent, respectively), while 0.6 percent of civil servants aged up to 30 were employed in B-category highest and principal public positions.

Among civil servants employed in B-category highest and principal public positions, the age group up of to 30 was most widely represented in administrations of federal courts and prosecutors' offices of the Russian Federation, where over a third of civil servants belonged to this age group.

Most civil servants employed in public positions (29.9 percent) had a length of service working in state bodies ranging from 5 to 10 years, while approximately 25 percent had a length of service in excess of 15 years. 42 percent of those employed in public positions in administrations of legislative bodies, 24 percent of those employed in executive bodies, judicial bodies and prosecutors' offices, and 38 percent of those employed in other state bodies had a length of service in excess of 15 years.

Thus, within the next 10 years over half of civil servants employed in key positions will retire, whereas it will be impossible to fill these positions with civil servants of a next generation with sufficient track record working in the civil service and possessing necessary professional skills.

The educational level of civil servants is characterized by the following parameters.

73 percent of civil servants had higher professional education at the beginning of 2002, with 40 percent of civil servants having higher economic and management education, 22 percent having legal education, and 18 percent having higher technical education. To compare, at January 1, 1999 civil servants having higher professional education in these areas accounted for 35.4, 19.6, and 21.2 percent, respectively.

The number of civil servants whose professional education does not match qualification requirements of their positions tends to grow.

Thus, 26.7 thousand people, or 4.6 percent of all civil servants, who are employed in positions that are often of a support nature have no professional education.

Retraining and professional development of federal civil servants is currently being funded from the federal budget. Under Federal Law On the Federal Budget for 2002, RUR 417.5 million was appropriated for this purpose, including RUR 68.5 million allocated for the state order that serves as a basis for retraining and professional development of federal civil servants of central administrations of federal executive bodies. At the same time, the federal budget does not provide funds for retraining and professional development of civil servants of territorial bodies of federal executive bodies.

The average accrued monthly salary (including base compensation and social benefits) of civil servants at all levels amounted to RUR 4.8 thousand in 2001.

The average monthly salary level of civil servants of executive bodies of the constituent entities of the Russian Federation was 1.4 times higher in 2001 than the average monthly salary level of civil servants of territorial bodies of federal executive bodies.

Civil servants are underpaid in comparison to managers and specialists of management bodies in the non-state sector of the economy. A narrow differentiation in compensation levels of civil servants provides little incentive for them to improve their professional performance.

According to experts' estimates, civil servants employed in comparable-level positions are paid 10–15 times less than executives (CEOs and deputy CEOs), 7-10 times less than senior managers (division heads), 3-7 times less than managers (specialists, project managers), and 1.5-3 times less than administrators (project implementers) [in the non-state sector of the economy].

The vast discrepancy in compensation levels between civil and municipal servants on the one hand and employees in the non-state sector of the economy on the other hand hampers recruitment of skilled specialists and managers to the civil and municipal service, leads to outflow of civil servants into the non-state sector of the economy (particularly affecting occupations and age categories that are in strongest demand), resulting in reduced performance of state bodies and in impaired professional, gender, and age civil servant structure.

In the process of harmonizing the regulatory framework of the civil service with the existing social relations and new economic environment, it is envisaged to optimize the compensation of civil servants in order to bring its level closer to compensation levels of managers and specialists of management bodies in the non-state sector of the economy, thereby making the state a competitive employer and improving the professional, gender, and age public servant structure.

The existing state of the civil service leads to the emergence of the following negative development trends.

The professional civil servant structure is deteriorating due to low average salary levels and significant salary differences between various state bodies. There is a growing shortage of civil servants younger than 45 who are experienced in working within the modern economic environment. At the same time, the prestige of the civil service as a profession is declining. Civil servants with non-applicable education credentials continue to prevail. Men prevail in management positions, while women prevail in specialist positions. The shortage of skilled personnel is compounded with the existence of vacancies that are hard to fill primarily because of low pay of civil servants.

Joining the civil service by citizens and job (service) promotion of civil servants are contingent on personal attitudes of state body heads towards them rather than on their professional skills and abilities. Frequent replacement of state body heads often leads to numerous changes in the structure of these bodies and unwarranted firings of civil servants.

The quality of documents prepared is often substandard. The activities of state bodies are insufficiently regulated and barely controlled by the civic society.

Outdated moral standards and ideological controls that used to regulate the behavior of employees of state and government bodies and to prevent abuse, corruption, and arbitrariness in these bodies have ceased to apply. Neither new rules for official behavior (professional ethics) of civil servants nor legal mechanisms for implementation thereof have been developed so far.

Shortcomings in the area of staff training for the civil service and professional development of civil servants have not been eliminated. The existing educational institutions in general fail to provide the required level of professional training for civil servants.

Positive foreign experience in the field of civil service development is not applied to a sufficient extent.

In order to overcome the aforementioned negative trends, it is necessary to solve the following problems facing the civil service:

- Violation of the principle of uniformity of the civil service and the civil service administration system at the federal and regional levels;
- Lack of developed legislative mechanisms of interaction between the civil and municipal services;
- Contradictions and gaps in the legislation of the Russian Federation on the civil service;
- Mismatch of the social and legal status of civil servants and the degree of their responsibility;
- Low performance efficiency of state bodies;
- Insufficient use of advanced state government techniques, including those that are consistent with the task of implementing large-scale state programs and projects;
- Low efficiency of legal and organizational controls over the activities of state bodies by the civic society;
- Low efficiency of civil service personnel policy;
- Preservation of a conservative system of civil servant training and professional development;
- Insufficient civil service resources.

All of the above problems are closely interrelated and cannot be solved separately from one another. The civil service reform should be implemented within the framework of a balanced development concept and should not consist of one-off radical changes in the existing civil service. A consistent and gradual approach is the underlying principle of modern civil service reform.

In order to solve the aforementioned problems, in accordance with the Instruction of the President of the Russian Federation No. Пр-2331 of November 24, 2000, a Concept of the Reform of the Civil Service of the Russian Federation was developed and approved on August 15, 2001 under No. Пр-1496, serving as a basis for developing this Program.

2. Program goal, objectives, time frame, and main implementation stages

Program goal – improving the efficiency of the civil service as a whole and of its types and levels, optimizing civil servant maintenance costs, and developing civil service resources.

To achieve this goal, the following objectives need to be implemented:

- Creating conditions for optimal organizational and legal support of the civil service;
- Defining roles, powers, and responsibilities of civil servants based on job (service) descriptions;
- Implementing new techniques for planning, financing, stimulating, and assessing civil servants' activities, and using civil service system resources in a rational manner;
- Ensuring openness of the civil service for the benefit of civic society development and strengthening of the state;
- Applying efficient methods for selecting skilled personnel for the civil service and for assessing service performance of civil servants, as well as creating conditions for civil servant job (service) promotion;
- Implementing staff training programs for the civil service and professional development programs for civil servants;
- Implementing mechanisms for identifying and solving civil service-related conflicts of interest, as well as introducing a legal regulation of professional ethics of civil servants;
- Creating optimal material and technical conditions for efficient functioning of the federal civil service and performance by civil servants of their official (service) duties;
- Ensuring the development of a public service management system.

The Program is medium-term and shall be implemented over the years 2003–2005.

Program implementation stages:

2002 – Preparations for Program implementation;

2003 – 2005 – Creating a regulatory framework for the civil service system, implementing financial and economic measures, experiments and pilot projects aimed at applying new approaches to federal civil service organization, monitoring civil service reform, as well as applying positive outcomes of these experiments and pilot projects in the activities of state bodies and local self-government bodies.

3. System of measures envisaged by the Program

The Program envisages a set of specific measures listed in the Appendix and main areas in which these shall be implemented.

In order to achieve the objectives set out in the Program, measures in the following areas are envisaged:

- Improving the legislative and regulatory framework for civil service reform;
- Experimenting and implementing pilot projects aimed at applying new approaches to federal civil service organization and support of federal civil servants' activities;
- Improving training, retraining, and professional development of federal civil servants;
- Creating a material and technical basis for efficient functioning of the federal civil service;
- Forming a civil service management system.
- Improving the legislative and regulatory framework for civil service reform

The need for legislative regulation of civil service issues is stipulated in the Constitution of the Russian Federation. However, no comprehensive and integral legal framework for civil service support exists in the Russian Federation.

The legal regulation of the civil service in the Russian Federation is deficient in that it does not provide a complete legislative regulation of civil service relations.

Legal frameworks for various types of the federal civil service are subject to various regulations of the Russian Federation, which contradict to one another.

The constituent entities of the Russian Federation have enacted various civil service regulations, which in certain cases are at odds with the legislation of the Russian Federation.

Therefore, no comprehensive regulatory framework exists that would ensure a complete implementation of objectives and functions assigned to the civil service under the Constitution of the Russian Federation and generally recognized principles and norms of international law.

In order to establish a comprehensive civil service system, the need has arisen to legislate a uniform legal framework for it.

It is necessary to develop and enact priority federal laws and other regulations on the civil service system of the Russian Federation, on types of the civil service, on local self-government, and on municipal service.

These laws and regulations shall provide a uniform approach to legal regulation and organization of the civil and municipal service, improvement of compensation system for civil and municipal servants, openness of the civil service for the benefit of civic society development and strengthening of the state, implementation of advanced techniques for forecasting, financing, assessing, stimulating, and providing logistic support to the civil service, and civil service administration.

An expert study of the existing regulatory framework of the civil service is also required, in addition to harmonizing it with the existing social relations and new economic environment.

Experimenting and implementing pilot projects aimed at applying new approaches to federal civil service organization and support of federal civil servants' activities.

The Russian Federation retains conservative approaches to federal civil service organization, which fail to ensure efficient functioning of the federal civil service and appropriate quality of provision by civil servants of public services to citizens and organizations.

Improving federal civil service organization and providing support to federal civil servants' activities require the implementation of advanced management technologies, including new methods for planning and financing federal state bodies' activities and stimulating federal civil servants, as well as of a system of indicators and criteria for assessing activities of administrations of federal state bodies and divisions thereof.

Solving these tasks requires experimenting and implementing pilot projects involving the application of new approaches to federal civil service organization in selected federal state bodies. The monitoring of the results of these experiments and pilot projects will be used as a basis for implementing advanced management technologies in the civil service system as a whole.

As part of the Program implementation, the following experiments and pilot projects are envisaged:

- Testing new methods for planning federal state bodies' activities, and a system of indicators and performance assessment criteria for federal civil servants;
- Implementing methods of targeted program financing of selected federal state bodies aimed at improving efficiency of these bodies and federal civil servants;
- Implementing the balance method of resource accounting and advanced forms of federal state bodies' budget spending efficiency assessment;
- Applying service contracts;
- Creating and using a system for federal civil servant professional development, training and retraining;
- Developing mechanisms for identifying and solving civil service-related conflicts of interest, as well as introducing a legal regulation of professional ethics of civil servants.
- Improving training, retraining, and professional development of federal civil servants

The main goal of improving training, retraining, and professional development of federal civil servants is to raise their professional level.

In order to achieve this goal, the following is required:

- Developing an efficient mechanism for selecting personnel for the civil service and for job (service) promotion of federal civil servants on the basis of their professional achievements and capabilities;
- Raising the prestige of the federal civil service and the authority of federal civil servants;
- Establishing a system of continuous professional education of federal civil servants, improving standards and training and professional development

programs for federal civil service employees on the basis of open tenders for such programs, as well as educational institutions;
- Managing the development of professional qualities of federal civil servants;
- Ensuring a renewal and scheduled rotation of federal civil service personnel;
- Forming a reserve of federal civil servants on a competitive basis and ensuring it is used efficiently;
- Assessing federal civil servants' performance in an objective manner at attestation or qualification examinations.
- Creating a material and technical environment for efficient functioning of the federal civil service

The main goal of this area is ensuring the use of civil service system resources in a rational manner, as well as creating an appropriate environment for high-quality performance by civil servants of their official (service) duties and provision by them of civil services to citizens and organizations.

In order to achieve this goal, the following objectives shall be implemented:

- Analyzing the existing material and technical base of the federal civil service;
- Developing workplace and accommodation standards for federal civil servants, as well as standards regulating logistical support of the federal civil service;
- Developing proposals on the construction, reconstruction, and technical outfitting of buildings and premises housing central administrations of federal state bodies.
- Forming a civil service management system

A civil service management system is required for ensuring the uniformity of the civil service of the Russian Federation, interaction of state bodies with civic society entities, coordination of state bodies' activities involving recruitment into the civil service, working in and termination from the civil service, as well as the professional development of civil servants.

This goal shall be achieved on the basis of:

- Uniformity of the administration system of the civil service of the Russian Federation and its division into the federal level and level of the constituent entities of the Russian Federation;
- Coordination of activities of state bodies' divisions involving civil service and personnel issues;
- Interaction with local self-government bodies;
- Openness of the civil service system for the citizens and civic society.

4. Program resources

The amount of Program funding in 2003–2005 shall be RUR 539.1 million, including:
in 2003 – RUR 231.7 million;
in 2004 – RUR 164 million;
in 2005 – RUR 143.4 million.

It is envisaged that federal budget funds appropriated for implementing the Program shall be allocated to the Administrative Department of the President of the Russian

Federation (to the Program Coordinator – Administration of the President of the Russian Federation) and subsequently transferred to the state customers.

Program implementation expenses shall be specified upon preparation of a draft federal budget for the relevant year.

In order to implement the measures envisaged by the Program, it will be possible to attract World Bank funds, as well as funds obtained via technical assistance programs provided by foreign countries and international organizations. Technical assistance may be used for advisory support of research and development measures envisaged by the Program in accordance with the legislation of the Russian Federation on foreign technical assistance.

Upon obtaining positive results of individual measures envisaged by the Program, funds required for the implementation thereof shall be allocated with account taken of capabilities of budgets at all levels for the relevant year.

5. Program implementation and monitoring progress in carrying out its measures

The Program shall be implemented using federal budget funds on the basis of state contracts (agreements) concluded according to the established procedure by state customers with implementers of the measures envisaged by the Program, and on the basis of agreements concluded with relevant constituent entities of the Russian Federation or municipalities.

Implementers of the measures envisaged by Program – Ministry of Labor and Social Development of the Russian Federation, Ministry of Trade and Economic Development of the Russian Federation, Ministry of Internal Affairs of the Russian Federation, Ministry of Defense of the Russian Federation, Ministry of Property Relations of the Russian Federation, Ministry of Finance of the Russian Federation, Ministry of Justice of the Russian Federation, State Committee for Construction and Housing of the Russian Federation, research and development, educational, and other entities to be determined on a tender basis according to the established procedure.

The Program Coordinator shall be responsible for organizing the Program implementation and monitoring progress in carrying out the measures envisaged therein.

The Program Coordinator shall ensure the implementation of the state policy in the area of civil service reform by coordinating approved federal and regional programs related to civil service development, and ensuring interaction between state bodies and local self-government bodies in implementing the Program.

The Program Coordinator shall review materials pertaining to the Program implementation progress, specify, on an annual basis, measures envisaged by the Program and related funding amounts, and approve a detailed list of measures to be implemented.

In order to ensure the Program implementation, state customers shall submit to the Program Coordinator their proposals and substantiation on the most urgent measures to be financed in the following financial year, as well as proposals concerning the use of funds obtained via foreign technical assistance.

State customers shall be responsible for developing terms of reference for competitions and tenders for implementing the measures envisaged by the Program. Tender committees formed according to the established procedure shall perform the tender selection.

The Program Coordinator shall be responsible for monitoring the Program implementation.

6. Program implementation deliverables

As a result of the Program implementation, the following results shall be achieved:

- Harmonizing the regulatory framework of the civil service with the existing social relations and new economic environment;
- Establishing a professional civil service based on the principles of taking into account and assessing civil servants' performance;
- Implementing job (service) descriptions for civil servants;
- Achieving a high-quality level of performance by civil servants of their official (service) duties and provision by them of civil services to citizens and organizations;
- Creating conditions for openness and accountability of state bodies' administrations and civil servants to the civic society;
- Creating a framework for improving financial and economic support of the federal civil service;
- Improving the efficiency of personnel policy within the civil service system in order to improve the quality of civil service employee structure;
- Implementing mechanisms for identifying and solving civil service-related conflicts of interest, as well as introducing a legal regulation of professional ethics of civil servants;
- Raising the professional level of federal civil servants;
- Optimizing the number of federal civil servants
- Creating a material and technical environment for efficient performance by civil servants of their official (service) duties;
- Establishing a civil service management system.

Concept and Main Directions of Reform and Modernization of Central State Administration

Extracts form a strategic document approved by the Czech Government in its Resolution N 237 of 17 March 2004[1]

Vision of the Reform

Flexible and better functioning central state administration responding better to global challenges and cross-cutting issues.

Why Central State Administration Reform?

The aim of reform is to ensure, through intervention into the current system, better enforcement of law and more effective services provided by the central state administration serving the Government and society. Central state administration authorities should be focused more on the quality of their services and outputs they provide and react in a better and more flexible way on demand expressed by citizens. Tools providing for systematic use of cost-benefit analysis on public finance to economize central state administration operational expenditure should be introduced. Implementation of Government policies should be enhanced, as well as background information submitted to the Government for its decision-making. The reform should also support wider use of modern technologies and managerial techniques and enhance horizontal co-ordination and communication among central state administration authorities.

Project management tool is being used for the implementation of the whole reform process.

Why Project Management?

Exploitation of the project management is appropriate mainly because of the following attributes:

- higher flexibility
- suppression of departments´ own preferences
- better targets orientation
- greater team co-operation
- greater transparency
- increased responsibility for outputs

Project management consists of five main activities with their internal order of sequence, however, these activities do not have to be always implemented separately:

- project targets definition
- planning
- management

[1] Source: www.vlada.cz

- monitoring
- project termination

Reform Co-ordination
The main responsibility for the reform falls within the duties of the Prime Minister.
Co-ordination of activities within project management and respective tasks in the reform falls within duties of the Head of the Office of the Government. For this purpose, Department of Regulatory Reform and Central State Administration Reform within the Office of the Government is charged with implementation of particular tasks.
Top level operational body of the reform is the Central State Administration Reform Steering Committee which monitors work on the reform and approves its further progress. All project outputs are submitted to this Steering Committee in order to express its opinion on it.
Project teams are established to provide implementation of separate projects headed by appointed project managers. Practical implementation of tasks is assigned to project teams which are created to each project. A project manager is appointed as a head of each project team as the responsible person for the progress of a respective project. Among others, to his / her duty belongs e.g. keeping the records of project documentation.

DIRECTIONS OF REFORM

A. Rationalization of processes in central state administration

Project A.1 Identification of the mission statement (and targets) of central state administration authorities
The purpose of this project is to define mission statements of all central state administration authorities and determine their responsibility for social values. In accordance with the mission statements and priorities of the government, departments will define their own mid-term priorities and targets. On the basis of submitted materials, the Office of the Government of the Czech Republic elaborates the summary document which will be submitted to the Government for discussion.
This project has a key role for further continuation of the reform towards rationalization of processes in the central state administration. On the basis of the outputs of this project current processes in central state administration will be defined and described. Subsequently, the rationalization of processes with the aim to increase effectiveness and quality of services in the central state administration will follow.
Time framework - till 31 December 2004
Output - Mission statements and mid-term targets of central state administration authorities
Responsible authority - Office of the Government - Department of Regulatory Reform and Central State Administration Reform
Co-responsible authorities - all central state administration authorities
Project Manager - Zuzana Rusňáková

Project A.2 Process description and analysis in central state administration authorities
The aim of the project is to implement process audits in all central administration authorities. Detailed mapping and analysis of current processes will enhance definition of those ones which have a key role for fulfillment of the mission statement and mid-term targets of a particular central administration authority, and their distinguishing from those identified as duplicitous, unproductive, or having nothing in common with the mission of such authority.
Time framework - till 31 December 2006
Input - outputs of the project A.1 Identification of the mission statement (and targets) of central state administration authorities
Outputs

1. detailed process description implemented in central state adminitration authorities
2. definition of "candidates" for transfer

Responsible authority - Office of the Government - Department of Regulatory Reform and Central State Administration Reform
Co-responsible authorities - all central state administration authorities
Project Manager - Aleš Pecka

Project A.3 Elaboration of general rules of operation for agencies
The aim of this project is to define common standard rules of the way how services to central administration authorities should be provided. It concerns those services which are not related to the mission statements of the central state administration authorities and are of supporting and operational nature. The fact that elaboration of the above-mentioned rule does not necessarily have to lead to creation of e new legal entity is important. Unification of existing conditions for establishment and operation of current bodies ensuring this kind of services to the central state administration should be the target of this project.
Time framework - till 30 June 2005
Output - Rules of operation for so-called "agencies"
Responsible authority - Office of the Government - Department of Regulatory Reform and Central State Administration Reform
Co-responsible authority - Ministry of the Interior, Ministry of Finance
Project Manager - Štěpánka Steinbachová

Project A.4 Reorganization of central state administration
The aim of this project is to set the rules for the "new" central state administration functioning, as well as embodying into the legislation the previous solutions which proved to be effective. The project should describe a final solution and provide the Government with background materials for its final decision.
Inputs

1. Output of the project A.1 Identification of the mission statement (and targets) of central state administration authorities
2. Output of the project A.2 Process description and analysis in central state administration authorities

3. Output of the project A.3 Elaboration of the general rules of operation for agencies

Time framework and outputs

1. Proposal for launching of operation of agencies - till 30 June 2006
2. New Draft Competence Law - till 30 June 2006
3. New central state administration and effectively performing agencies - till 1 January 2007

Responsible authority - Office of the Government - Department of Regulatory Reform and Central State Administration Reform

Co-responsible authority - all central state administration authorities

Project Manager - Luděk Tesař

B. Improvement of management in central state administration

Project B.1 Effective horizontal communication and support of nationwide strategies

Balance between a certain level of autonomy of central state administration authorities on the one hand, and demand for their co-ordinated operation on the other, is a necessary precondition for effective performance of the central state administration as a whole. Definition of rules for horizontal communication among central state administration authorities, and reduction of the inter-departmental administrative work as well as decrease of ′communication overload′ of ministries and Government is, therefore, the aim of this project.

Time framework - till 31 December 2005

Input - Analysis of foreign experiences with the inter-departmental coordination and functioning of the offices of the government or prime minister

Outputs

1. Draft rules and institutional framework for effective horizontal communication
2. Elaboration of the system supporting the creation of nationwide strategies and cross-cutting solutions

Responsible authority - Office of the Government - Department of Regulatory Reform and Central State Administration Reform

Co-responsible authority - Ministry of the Interior

Project Manager - Martina Jalovecká

Project B.2 Modern managerial techniques in central state administration

This project is focused on implementation of a training system providing for a support and creation of a skilled top management in central state administration authorities capable of use of modern managerial tools. For this purpose, managerial capacities will be systematically built and promoted in central authorities with the aim to introduce and further disseminate the skills of e.g. performance management; strategic management; performance (output) focused budgeting; human resources management etc.

Time framework and outputs

- Preparation and implementation of a pilot project on output focused management - till 31 December 2004

- Implementation of the training project form managers focused on output focused management - till 31 December 2005
- Processing of the Strategic paper on the development of management in central state administration authorities for the period 2006 till 2010 - till 31 December 2005

Responsible authority - Office of the Government - Directorate General of the Civil Service
Project Manager - Zuzana Beránková

Project B.3 Better co-ordination of central state administration towards territorial public administration
The aim of this project is an identification of optimal co-ordination mechanism of communication towards territorial public administration and improvement of co-ordination and communication interface between central administration authorities and other entities in the public sector (municipalities, business sector, associations etc.).
Time framework and outputs

- Analysis of co-ordination mechanisms of central state administration towards territorial public administration - till 31 December 2004
- Proposal of the co-coordinated approach of central state administration towards territorial public administration - till 31 December 2005

Responsible authority - Ministry of the Interior
Co-responsible authority - Office of the Government - Department of Regulatory Reform and Central State Administration Reform, Ministry of Informatics
Project Manager - Jaroslav Maršík

Project B.4 Knowledge management
Knowledge management is mainly related to acquiring, processing, transmission, recording and sharing of knowledge. However, this issue does not concern only technical aspects. The project will mainly follow the outputs of information audits realized within the framework of the project A.2. The stress will be put mostly on information sharing among central state administration authorities, as well as keeping the "institution memory" - knowledge management within those authorities despite existing high fluctuation of employees.
Input

- Output of the project A.2 Process description and analysis in central state administration bodies
- OECD documents
- Strategy of the development of human resources in the Czech Republic
- Time framework and outputs
- Specification concerning the intent of knowledge management system adapted in relation to parameters and assumptions for use of the outputs from the process analysis - til 31 December 2004
- Launching of a pilot project to verify the knowledge management model - till 31 December 2006
- Creation of the knowledge management system - till 31 December 2007

Responsible authority - Office of the Government
Co-responsible authority - Office of the Government - Directorate General of the Civil Service, Ministry of Informatics

Project Manager - Jaroslav Hrubý

C. Improvement of quality in central state administration

Project C.1 Introduction and development of quality management in central state administration

The aim of the first stage of this project is to train management of central state administration authorities in quality management issues. This will enable general application of chosen quality method in central state administration authorities. The second stage shall consist of pilot project implementation of quality monitoring, application of CAF (Common Assessment Framework), as well as benchmarking method.

Input - output of the project A.2 Process description and analysis in central state administration authorities

Time framework and outputs

- Initiation of training, support and creation of quality management in central state administration - 2005 to 2006
- Pilot project on introduction of quality monitoring into central state administration - 2007
- Application of CAF model benchmarking method into central state administration - 2008 to 2010

Responsible authority - Office of the Government - Department of Regulatory Reform and Central State Administration Reform

Projekt Manager - Štěpánka Steinbachová

Project C.2 Regulatory reform in central state administration

The process of modernizing regulation should contribute to enhancing preparation and implementation of laws and by-laws issued by state administration. Primary tool for better regulation agenda in the Czech Republic will be the implementation of Regulatory Impact Assessment (RIA) to the legislation-making process.

Time framework and outputs

- Draft rules for functioning of state regulatory authorities (Unified Model of State Regulatory Authorities) - till 31 December 2004
- Draft guidelines for Regulatory Impact Assessment (RIA) - do 31. December 2005
- Pilot project of RIA implementation on preparation of selected draft laws - 2006 to 2007
- Implementation of full RIA for all Government draft laws - since 2007

Responsible authority - Office of the Government - Department of Regulatory Reform and Central State Administration Reform

Project Manager - Daniel Trnka

Project C.3 E-government

The central state administration modernization should be supported by promoting the e-government tools. The aim of e-government implementation is to enable citizens and businesses to simply and quickly communicate with administrative authorities on all levels of public administration.

Time framework and outputs

- Preparation of project specification and launch of the second phase of the Public Administration Portal - till 31 December 2004
- Creation of basic exchange format for communication between public administration authorities - till 31 December 2004
- Implementation of basic electronic on-line public service through the Public Administration Portal - till 31 December 2005
- Implementation of basic exchange format for communication between public administration authorities - till 31 December 2006
- Creation of basic exchange format for communication between citizens and public administration authorities - 2004 to 2007

Responsible authority - Office of the Government - Department of Regulatory Reform and Central State Administration Reform
Project Manager - Daniel Trnka

D. Implementation and improvement of civil service

Project D.1 Civil service Act implementation
Human resources management in central state administration has not been unified yet. The implementation of the Act No. 218/2002 Coll., on service of civil servants in administrative authorities and on remuneration of civil servants and other employees in administrative authorities (The Civil Service Act) will be solved within the framework of the project. The project will progress in line with the Civil Service Act and priorities chosen for the period before the Act come fully into force.
Time framework and outputs

- Creation of personal departments according to the Civil Service Act - till 31 December 2004
- Preparation of draft systemization and organizational structure of administrative authorities for the year 2005 - till 1 September 2004
- Preparation of the Institute of State Administration transformation into a state contributory organisation according to the Civil Service Act - till 30 September 2004
- Issue of service regulation by the Director General of the Civil Service - till 31 December 2004
- Creation of guidelines for training of civil servants and candidates - till 31 December 2004
- Creation of the plan of supervisory activities for Directorate General of the Civil Service for the year 2005 - till 31 December 2004
- Creation of principles for the service appraisal of civil servants - till 31 December 2005

Responsible authority - Office of the Government - Directorate General of the Civil Service
Project Manager - Ivana Nemeškalová

E. Rationalisation of the finance and budgetary procedures in central state administration

Project E.1 Development of financial and performance management

The aim of the project is to prepare top management in central state administration for wide economic reforms. That particularly means to provide them with knowledge and skills enabling them to create a strong expert backround for further changes in economic management of central state administration. The Institute of State Administration will ensure realisation of educational and training programmes according to the content prepared by the responsible authority (MoF).
Time framework and outputs

- Plan on promotion and creation of performance and financial management in central state administration - till 31 December 2004
- Performance and financial management in central state administration - 2006 to 2008

Responsible authority - Ministry of Finance
Co-responsible authority - Office of the Government - Directorate General of the Civil Service
Project Manager - Bohdan Hejduk

Project E.2 Public Private Partnership
There are many investment projects which are of the interest of businesess as well as public sector. The aim is to find the rules for using both financial resources - public and private sector - in such a way that the businesses would be motivated to take part in public investments and to initiate them and also to guarantee that public resources will not be abused.
Time framework and outputs

- Draft guidelines for common public / private investment - till 31 December 2005
- Pilot projects on public investments and their evaluation - 2006 to 2010

Responsible authority - Ministry of Finance
Co-responsible authorities
Ministry of the Interior, Office of the Government - Department of Regulatory Reform and Central State Administration Reform, Ministry of Industry and Trade
Project Manager - Kateřina Helikarová

Project E.3 Unification and enhancement of control in central state administration
The project focuses on control in central state administration from the systematic point of view. The aim is to unify control into a wider framework and to promote the links among control, effectiveness and performance of central state administration authorities, in connection with their missions and mid-term targets. The issue of the body or department in charge of the complex policy co-ordination and methodical management of control will be solved as well.
Time framework and outputs

- Analysis of the current system of control in state administration - till 31 December 2004
- Strategy of control in state administration with focus on control of targets (objectives) and performance of central state administration authorities - till 31 December 2004
- Identification of the body co-ordinating and methodically managing control in state administration as a whole - till 31 December 2005

Responsible authority - Ministry of Finance
Co-responsible authority - Office of the Government - Directorate General of the Civil Service
Ministry of the Interior
Project Manager - Josef Svoboda

Standardized Interview: Questions and Responses of Russian and Czech Civil Servants

A standardized structured interview of the Czech and Russian civil servants took place in the Czech Republic in 2002 and in Russia in 2004. The interview was carried out in the Czech Republic by V. King and her team (King, 2003)[1] as part of the research project funded by CERGE-EI Foundation (Prague). In the Russian Federation the interview was conducted by the author of this study using the same questions. All in all 19 Russian and 19 Czech senior and middle level civil servants were interviewed, who represented various government ministries and agencies.

Provided below are the responses to 24 questions put to the interviewees:

1. What is the lowest position status, in which people in your ministry sign important letters in their own name?	Russian Federation	Czech Republic
Minister or his Deputy	5	
Director of Department	2	3
Deputy Director of Department	11	
Non-respondents	1	16

2. Do you find decision-making at your ministry too centralized, too decentralized, or just right?	Russian Federation	Czech Republic
Too centralized	6	4
Too decentralized	1	-
Just right	12	15

3. Do you have enough creative space?	Russian Federation	Czech Republic
Yes	13	19
No	6	-

[1] King V., (2003), 'What Kind of Civil Service? Trends in Public Administration Reform in Eastern Europe', in *NEBI Yearbook 2003*, Springer., pp. 55-66

4. Do you feel too restricted by rules?	Russian Federation	Czech Republic
Yes	6	-
No	13	19

5. Do you want more rules?	Russian Federation	Czech Republic
Yes	14	13
No	5	6

6. Do you feel burdened by having to make too many decisions on your own?	Russian Federation	Czech Republic
Yes	1	-
No	18	19

7. Is there a strong sense of teamwork at your ministry or do people act individualistically?	Russian Federation	Czech Republic
Strong teamwork	3	13
Weak teamwork	16	6

8. Is it easy to find qualified employees for your ministry (department)?	Russian Federation	Czech Republic
Yes	1	7
No	18	12

9. What motivates young people to come and work at your ministry? *In brackets ranked as most important - (1), ranked 2nd - (2), ranked 3rd -(3)*	Russian Federation	Czech Republic
Opportunity to make good	14 (1); 4 (2); 1 (3)	15 (1); 2(2); 2 (3)

contacts		
Challenge of the task	1 (2); 7 (3)	3 (1); 9 (2); 3 (3)
Status	2 (1); 9 (2); 3 (3)	1 (1); 3 (2); 8 (3)
Policy design	3 (3)	2 (2); 1 (3)
Job security	3 (1); 5 (2); 2 (3)	5 (3)
Could not find another job	3 (3)	3 (2)
Attractive pay		

10. What motivates employees to stay at your ministry for a longer time? *In brackets ranked as most important - (1), ranked 2nd - (2), ranked 3rd -(3)*	Russian Federation	Czech Republic
Long term career opportunity	10 (1); 4 (2); 5 (3)	18 (1); 1 (3)
Challenge of the task	3 (1); 3 (2); 2 (3)	1 (1); 5 (2); 3 (3)
Status	5 (2); 7 (3)	10 (2); 3 (3)
Policy design	3 (3)	1 (2); 3 (3)
Job security	6 (1); 6 (2);	7 (3)
Loyalty to the group		3 (2); 2 (3)
Attractive pay	1 (2); 2 (3)	

11. What kinds of things give you the most professional satisfaction? *In brackets ranked as most important - (1), ranked 2nd - (2), ranked 3rd -(3)*	Russian Federation	Czech Republic
Creative design of policy	9 (1); 4 (2); 6 (3)	18 (1); 1 (3)
Respect of the public and colleagues	9 (1); 5 (2); 1 (3)	1 (1); 5(2); 3 (3)
Feeling of authority	2 (2); 6 (3)	10 (2); 3 (3)
Providing good service	8 (2); 6 (3)	1 (2); 3 (3)

12. How do you evaluate the quality of a completed task or the quality of a performance? *In brackets ranked as most important - (1), ranked 2nd - (2), ranked 3rd -(3)*	Russian Federation	Czech Republic
How well accepted by the public at large	6 (1); 9 (2); 4 (3)	12 (1); 5 (2) 2 (3)

True to plan	12 (1); 7 (2); 2 (3)	6 (1); 1(2);
What percent of the target group did it reach	1 (1); 3 (2); 11 (3)	1 (1) 7(2); 3 (3)
Cost effectiveness	2 (3)	3 (2) 5 (3)

13. Are there overlapping jurisdictions between your ministry and other ministries?	Russian Federation	Czech Republic
Yes	18	19
No	1	-

14. Should there be more coordination between the activities of ministries and government agencies in this country?	Russian Federation	Czech Republic
Yes	18	11
No	1	8

15. Do you think that your country in general has the right mix of what the public sector does and what the private sector does?	Russian Federation	Czech Republic
Yes	7	18
No	12	1

16. Should there be a strict separation between Civil Service and Politics?	Russian Federation	Czech Republic
Yes	17	19
No	2	-

17. In your country civil service and politics are kept separate?	Russian Federation	Czech Republic
Yes	2	-
No	17	19

18. In your country change in Government entails replacement of some career civil servants?	Russian Federation	Czech Republic
Yes	19	19
No	-	-

19. Has the organization of your ministry been designed by following a model of some particular country?	Russian Federation	Czech Republic
Yes	4	8
No	15	11

20. Is it more important for your ministry (department) to achieve results or to guarantee the integrity of important on-going processes?	Russian Federation	Czech Republic
Results more important	5	5
Integrity of processes more important	4	7
Both are important	10	7

21. Is there sufficient dialog between policy makers and those affected by the policies ?	Russian Federation	Czech Republic
Yes	1	1
No	14	15
No response	4	3

22. How important is shielding of civil service from direct political influence?	Russian Federation	Czech Republic
Very important	16	19
Important	3	-
Not important	-	-

23. How important is a culture of civil service impartial professionalism?	Russian Federation	Czech Republic
Very important	18	18

Important	1	1
Not important	-	-

24. How important is the attainment of administrative stability?	Russian Federation	Czech Republic
Very important	14	17
Important	5	2
Not important	-	-

Printed in Dunstable, United Kingdom

91347965R00163